# Lecture Notes in Mathematics

Giuseppe Buttazzo · Aldo Pratelli
Sergio Solimini · Eugene Stepanov

# Optimal Urban Networks via Mass Transportation

Springer

Giuseppe Buttazzo
Dipartimento di Matematica
Università di Pisa
Largo Bruno Pontecorvo 5
56127 Pisa
Italy
buttazzo@dm.unipi.it

Aldo Pratelli
Dipartimento di Matematica
Università di Pavia
Via Ferrata 1,
27100 Pavia
Italy
aldo.pratelli@unipv.it

Sergio Solimini
Dipartimento di Matematica
Politecnico di Bari
Via Amendola 126/b
70126 Bari
Italy
solimini@dm.uniba.it

Eugene Stepanov
St. Petersburg University
of Information Technology, Mechanics
and Optics
Kronverkskij pr. 49
197101 St. Petersburg
Russia
stepanov.eugene@gmail.com

ISBN: 978-3-540-85798-3 e-ISBN: 978-3-540-85799-0
DOI: 10.1007/978-3-540-85799-0

Lecture Notes in Mathematics ISSN print edition: 0075-8434
ISSN electronic edition: 1617-9692

Library of Congress Control Number: 2008935629

Mathematics Subject Classification (2000): 49J45, 49Q10, 49Q15, 49Q20, 90B06, 90B10, 90B20



*Cover design*: SPi Publisher Services

Printed on acid-free paper

9 8 7 6 5 4 3 2 1

springer.com

# Preface

The monograph is dedicated to a class of models of optimization of transportation networks (urban traffic networks or networks of railroads and highways) in the given geographic area. One assumes that the data on distributions of population and of services/workplaces (i.e. sources and sinks of the network) as well as the costs of movement with and without the help of the network to be constructed, are known. Further, the models take into consideration both the cost of everyday movement of the population and the cost of construction and maintenance of the network, the latter being determined by a given function on the total length of the network. The above data suffice, if one considers optimization in long-term prospective, while for the short-term optimization one also needs to know the transport plan of everyday movements of the population (i.e. the information on "who goes where"). Similar models can also be adapted for the optimization of networks of different nature, like telecommunication, pipeline or drainage networks. In the monograph we study the most general problem settings, namely, when neither the shape nor even the topology of the network to be constructed is a priori known.

To be more precise, given a region $\Omega \subseteq \mathbb{R}^N$, we will model the transportation network to be constructed by an a priori generic Borel set $\Sigma \subseteq \Omega$. We consider then the mass transportation problem in which the paths inside and outside the network $\Sigma$ are charged differently. The aim is to find the best location for $\Sigma$, in order to minimize a suitable cost functional $\mathfrak{F}(\Sigma)$, which is given by the sum of the cost of transportation of the population, and the penalization term depending on the length of the network, which represents the cost of construction and maintenance of the network. To study the problem of existence of optimal solutions, we present first a relaxed version of the optimization problem, where the network is represented by a Borel measure rather than a set, and we prove the existence of a relaxed solution. We will study then the properties of optimal relaxed solutions (measures) and prove that, under suitable assumptions, the relaxed solution solve the original problem, i.e. in fact they correspond to rectifiable sets, and therefore can be called classical solutions. However, it will be shown that in general the problem studied may have no classical solutions. We will also study some topological properties of optimal networks, like closedness and the number of connected components. In particular, we find rather sharp conditions on problem data, which ensure the existence of closed optimal networks and/or optimal networks having at most countably many connected components. Finally, we will prove a general regularity result on optimal networks. Namely, we will show that an optimal network is covered by a finite number of Lipschitz curves of uniformly bounded length, although it may have even uncountably many connected components.

# Acknowledgments

This work was conceived during the meeting *Giornate di Lavoro in "Calculus of Variations and Geometric Measure Theory"* held in Levico Terme (Italy), and was carried on thanks to the project "Calcolo delle Variazioni" (PRIN 2004) of the Italian Ministry of Education. The work of the third author was partially supported by the italian GNAMPA–INDAM.

# Contents

# Chapter 1
# Introduction

The present monograph treats one particular class of mathematical models arising in urban planning, namely, the models of optimization of transportation networks such as urban traffic networks, networks of tram or metro lines, railroads or highways. The optimization is performed so as to take into account the known data of the distributions of the population and of services/workplaces (or, more generally, sources and sinks of the network), the costs of the transportation with and without using the network to be constructed, and the budgetary restrictions on construction and maintenance of the network, as well as, in certain cases, the transportation plan of everyday movement of the population. As an illustration, see the distribution of population as well as the railroad network in Italy (Figure 1.1). The functional to be minimized corresponds to the overall cost of everyday transportation of population from their homes to the services together with the cost of construction and maintenance of the network. It is important to emphasize that the shape and even the topology of the network is considered a priori unknown.

From the most general point of view such models belong to the class of economical optimal resource planning problems which were first studied in [44]. In the simplest cases under additional restrictions on the network such problems reduce to problems of minimization of so called average distance functionals (see [20]), and are similar to the well-known discrete problems of optimization of service locations (so-called Fermat-Weber, or $k$-median problems) studied by many authors (see, e.g. [7, 68, 69, 51]). Similar as well as slightly different models have been proposed for telecommunication, pipeline and drainage networks in [11, 41, 47], and are recently subject to extensive study (see, for instance, [8, 9, 10, 17, 27, 34, 48, 55, 56, 62, 66, 52, 73, 74]. The common kernel of all such models is the general (i.e. not necessarily discrete) setting of the Monge-Kantorovich optimal mass transportation problem (see, e.g. [42, 43, 67, 1, 36, 35, 60, 25, 38]); we give now a short description of the mass transport problem, a more complete discussion is given in Appendix A.

G. Buttazzo et al., *Optimal Urban Networks via Mass Transportation*,
Lecture Notes in Mathematics 1961, DOI: 10.1007/978-3-540-85799-0_1,

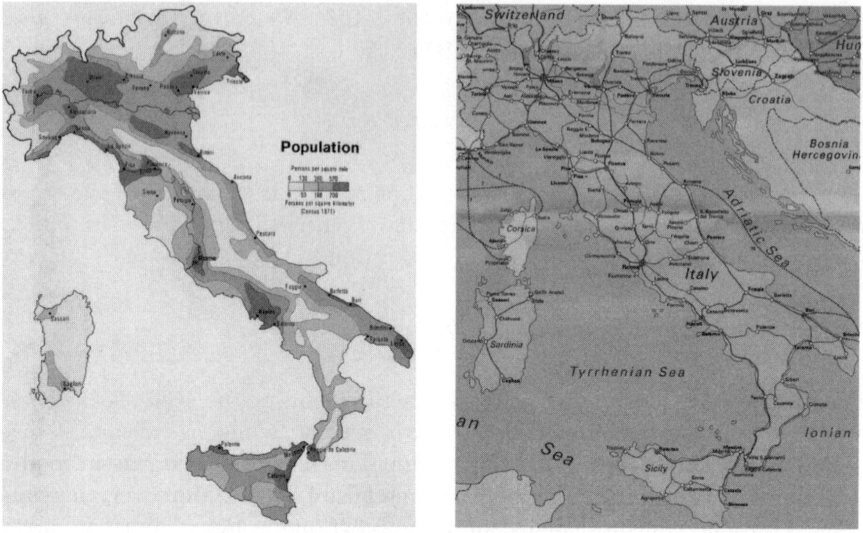


**Fig. 1.1** Density of population (left) and railway network (right) in Italy

The mass transportation problem was first proposed by Monge [49]. Using a modern language, this can be restated as follows: we are given a metric space $(X, d)$ and two finite Borel measures $f^+$ and $f^-$ with the same total mass $\|f^+\| = \|f^-\|$. A Borel map $T : X \to X$ is said to be a *transport map* if it moves $f^+$ on $f^-$, that is, if $T_\# f^+ = f^-$ being $T_\#$ the push-forward operator (see Appendix B.2). We are also given the *cost function*, which is a lower semicontinuous function $c : X \times X \to \mathbb{R}^+$; its meaning is very simple, namely $c(x, y)$ is the cost to move a unit mass from $x$ to $y$. In the original setting of Monge $c(x, y) = d(x, y)$, more generally one is often interested in $c(x, y) = d(x, y)^p$. The Monge transport problem consists then in determining, among all the transport maps, the *optimal transport maps*, that is, those maps which minimize the total transportation cost given by

$$\int_X c\big(x, T(x)\big)\, df^+(x)\,.$$

It may easily happen that there are no transport maps at all, namely when the measure $f^+$ has singular parts; it may also happen that, even thought there are transport maps, the existence of optimal transport maps fails. Also for this reason, it reveals of primary importance to consider the relaxed form of the problem proposed by Kantorovich (see [42, 43]). The idea of Kantorovich is to define *transport plan* any positive measure $\gamma$ on $X \times X$ such that the two marginals of $\gamma$ are precisely $f^+$ and $f^-$; the meaning is quite intuitive: such a measure $\gamma$ is to be interpreted as the strategy of transportation which moves a mass $\gamma\big(\{(x, y)\}\big)$ from $x$ to $y$; more precisely, it moves a total amount

$\gamma(C \times D)$ of mass from the set $C$ to the set $D$. An *optimal transport plan*, then, is any transport plan $\gamma$ minimizing the cost

$$\iint_{X\times X} c(x,y)\, d\gamma(x,y)\,.$$

It is to be noticed that the transport plans are a generalization of the transport maps: indeed, given a transport map $T$ the measure $\gamma_T := (\mathrm{Id}, T)_\# f^+$ is a transport plan, and moreover by definition

$$\iint_{X\times X} c(x,y)\, d\gamma_T(x,y) = \int_X c\big(x, T(x)\big)\, df^+(x)\,;$$

so, the search of optimal plans is a generalization of the search of optimal maps. The power of this new definition is evident: while, as we said, it may happen that there are no transport maps, or no optimal transport maps, there are always transport plans, as for instance $f^+ \otimes f^-$. Moreover, there are always optimal transport plans, since the function $c$ is lower semicontinuous. A more detailed introduction to mass transportation problems is given in Appendix A.

In this monograph we consider a problem of urban planning, in which we take as ambient space a region $\Omega \subseteq \mathbb{R}^N$, with $N \geq 2$ since the one-dimensional case is in fact trivial; the measure $f^+$ represents the density of the population in the urban area $\Omega$ and the measure $f^-$ represents the density of the services or workplaces. We also consider a Borel set $\Sigma \subseteq \Omega$ of finite $\mathscr{H}^1$ length, which represents the urban transportation network that has to be constructed to minimize the cost of transporting $f^+$ on $f^-$ according to some suitable cost functional.

Once the set $\Sigma$ is given, the cost $d_\Sigma(x,y)$ to be paid in order to connect any two points $x$ and $y$ of $\Omega$ is defined as the least "price" of moving along a Lipschitz curve connecting $x$ and $y$ given by the number

$$\delta_\Sigma(\theta) := A\big(\mathscr{H}^1(\theta \setminus \Sigma)\big) + B\big(\mathscr{H}^1(\theta \cap \Sigma)\big)\,.$$

The functions $A$ and $B$ are two given nondecreasing functions from $\mathbb{R}^+$ to $\mathbb{R}^+$ with $A(0) = B(0) = 0$, $A$ being continuous and $B$ lower semicontinuous: $A(s)$ is the "cost" of covering a distance $s$ by own means, that is a number including the expenses for the fuel, the fare of the highway, the fatigue of moving by feet, the time consumption and so on; on the other hand, $B(s)$ represents the cost of covering the distance $s$ making use of the transportation network (i.e. the "cost of the ticket").

In this monograph, we assume the point of view of an "ideal city", where the only goal is to minimize the total expenses for the people; therefore, the number $B(s)$ should be regarded just as a tax that people pay to contribute to the cost of the network when they use it, and the case $B \equiv 0$, corresponding to a situation where everybody can use the public transportation for free, is the simplest (and most common in the literature) choice in this ideal setting.

An opposite point of view, where the owner of the network aims to maximize his total income by choosing a suitable pricing policy $B$, has been studied in [18].

Having fixed the set $\Sigma$, the population will naturally try to minimize its expenses, that is, people choose to move following a transport plan $\gamma$ minimizing

$$I_\Sigma(\gamma) := \iint_{X\times X} d_\Sigma(x,y)\, d\gamma(x,y)$$

among all admissible transport plans, and we denote by $MK(\Sigma)$ the respective minimum (or the infimum if the minimum is not achieved). We want to find a network $\Sigma$ minimizing the total cost for the people. However, $MK(\Sigma)$ is not the only cost to be considered: otherwise, a network of infinite length covering the whole $\Omega$ would be clearly the optimal choice. We will then consider also a very general cost function $H\big(\mathscr{H}^1(\Sigma)\big)$ for the maintenance of the network, that will depend on the length $\mathscr{H}^1(\Sigma)$ of $\Sigma$ and that diverges if the length goes to $\infty$. For instance, one can set

$$H(l) := \begin{cases} 0, & \text{if } l \le L, \\ +\infty, & \text{if } l > L, \end{cases}$$

which corresponds to a situation where one is allowed to build a network of total length not exceeding $L$. Our goal is then to find an optimal network $\Sigma_{\rm opt}$ which minimizes $\Sigma \mapsto MK(\Sigma) + H\big(\mathscr{H}^1(\Sigma)\big)$ among the admissible sets $\Sigma$.

The above problem can be considered as a long-term optimization model. In fact, in this case while choosing the optimal network $\Sigma$ one is allowed to change freely the transportation plan $\gamma$ (i.e. it is supposed that people may consider it more convenient to choose different destinations for their everyday movements, e.g. change the shops they usually use or even change their workplace, in view of the cost of transportation), which is only reasonable in a quite long-term prospective. On the contrary, the reasonable model for the short-term prospective is obtained by considering given the transport plan $\gamma$ (i.e. the information on "who goes where" in the everyday movements) and thus minimizing $\Sigma \mapsto I_\Sigma(\gamma)+H\big(\mathscr{H}^1(\Sigma)\big)$ among the admissible sets $\Sigma$. However, it is easy to notice, similarly to [18], that the short-term optimization problem is in fact simpler than the long-term one. Hence in this monograph we concentrate on studying the latter with all the results applying also to the former.

## Plan of the Monograph

In Chapter 2 we define the general problem setting without additional assumptions on admissible networks. The simplest case, when $\Sigma$ is a priori required to be connected, will be considered in Chapter 3, and some known facts about this problem will be reported. In this case, by a suitable use of the Hausdorff convergence on connected sets, we show the existence of an optimal network. A particular situation happens when the goal of the planner is simply to transport the source mass $f^+$ to a network $\Sigma$ in the most efficient way, that is $f^-$, instead of being a priori fixed, is chosen in an optimal way among the probabilities with support in $\Sigma$. This problem then corresponds to the minimization of the functional

$$F(\Sigma) := \int_\Omega A\big(\mathrm{dist}\,(x,\Sigma)\big)\,df^+(x)\,. \tag{1.1}$$

We will refer to the minimization problem for the functional $F$ defined by (1.1) as the *irrigation problem* in view of the natural interpretation of the cost (1.1) as the total effort to irrigate the mass distribution $f^+$ using a network $\Sigma$. It is assumed that the effort to irrigate the point $x \in \Omega$ depends on its distance $t$ from the network $\Sigma$ through the function $A(t)$. Taking $A(t) := t$ we have the minimization problem for the *average distance functional*

$$\min\Big\{\int_\Omega \mathrm{dist}(x,\Sigma)\,df^+(x)\ :\ \Sigma\subseteq\Omega,\ \Sigma\text{ connected},\ \mathscr{H}^1(\Sigma)\le L\Big\},$$

that has been studied in several recent papers (see, e.g. [17, 21, 19, 20, 54]). On Fig. 1.2 below we show the plot of two cases when $\Omega$ is the unit bidimensional disc, $f^+$ is the Lebesgue measure over $\Omega$, and $\Sigma$ varies among all connected sets of length $L$, with two different choices of $L$.

It is immediate to see that dropping the connectedness assumption leaving the cost functional as in (1.1) would give zero as the minimal value of $F$, since the set $\Sigma$ would have the interest to spread everywhere on $\Omega$. This is why the particular situation considered by functional (1.1) is meaningful only in the connected framework.

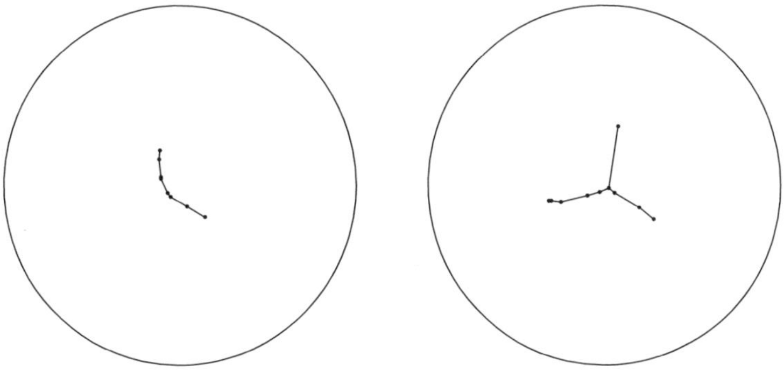

**Fig. 1.2** Optimal irrigation networks for $L = 0.5$ (left) and $L = 1$ (right)

In Chapter 4 we show that without extra assumptions on the functions $A$, $B$ and $H$ there may be no optimal networks. Therefore, we introduce a relaxed version of the problem, where the sets are replaced by Radon measures, and in particular each set $\Sigma$ corresponds to the measure $\mathscr{H}^1 \llcorner \Sigma$. Then, we show the existence of optimal "relaxed networks", and in particular we prove that optimal measures $\mu$ on $\Omega$ of the form $\mu = a(x)\mathscr{H}^1 \llcorner \Sigma$ for a one-dimensional rectifiable set $\Sigma$ and a Borel function $a : \Sigma \to [0,1]$ always exist. Roughly speaking, this means that there is an optimal transportation network concentrated on a Borel set $\Sigma$, but it has a pointwise density in $[0,1]$: the density 1 corresponds to a standard railway, where covering a path of length $l$ has a cost $B(l)$. In general, covering a path of length $l$ on a network of density $0 \leq p \leq 1$ costs $A\big((1-p)l\big) + B\big(pl\big)$, as if one covers a length $pl$ on the network, and the remaining $(1-p)l$ by own means. Moreover we show that, under suitable assumptions, there are also "classical solutions", that is, optimal networks which naturally correspond to sets (in other words, relaxed solutions with the coefficient $a(x)$ above taking only values 0 and 1). However, we give counterexamples showing that this does not always occur.

In Chapter 5 we consider two questions, namely whether or not there exists an optimal classical network which is closed, or which has only countably many connected components. We present counterexamples to show that this is not always the case, even when classical solutions exist. However, we are able to find conditions under which one has the existence of an optimal classical network that is closed or has countably many connected components.

In Chapter 6 we prove that, under suitable hypotheses, there is a classical optimal network that is covered by a finite number of Lipschitz curves of uniformly bounded length, even if it may still have infinitely many (even more than countably many) connected components.

Finally, the monograph is concluded by two appendices, which present with more details the general mass transportation problem and some tools from Geometric Measure Theory, among which the Disintegration Theorem and the $\Gamma$-convergence, which are used through the volume.

# Chapter 2
# Problem Setting

In this chapter we introduce the notation and the preliminaries to rigorously set the problem of optimal networks. The formulation in the sense of L. Kantorovich, by using *transport plans*, i.e. measures on the product space $\Omega \times \Omega$, will be presented together with a second equivalent formulation where the main tools are the so-called *transport path measures* that are measures on the family of curves in $\Omega$. This seems to be a very natural formulation that has already been used in previous papers (see for instance [24, 65, 6, 58]) and that allows to obtain in a rather simple way existence results and necessary conditions of optimality.

## 2.1 Notation and Preliminaries

In this monograph the ambient space $\Omega$ is assumed to be a bounded, closed, $N-$dimensional convex subset of $\mathbb{R}^N$, $N \geq 2$, equipped with the Euclidean distance; the convexity assumption is made here only for simplicity of presentation; in fact, all the results are still valid in the more general case of bounded Lipschitz domains. For any pair of Lipschitz paths $\theta_1$, $\theta_2 : [0,1] \to \Omega$, we introduce the distance

$$\begin{aligned} d_\Theta(\theta_1, \theta_2) := \inf \Big\{ &\max_{t\in[0,1]} |\theta_1(t) - \theta_2(\varphi(t))| \,, \\ &\varphi : [0,1] \to [0,1] \text{ increasing and bijective} \Big\} \,, \end{aligned} \tag{2.1}$$

where $|\cdot|$ is the Euclidean norm in $\mathbb{R}^N$. We define then $\Theta$ as the set of the equivalence classes of Lipschitz paths in $\Omega$ parametrized over $[0,1]$, where two paths $\theta_1$ and $\theta_2$ are considered equivalent whenever $d_\Theta(\theta_1, \theta_2) = 0$: it is easily noticed that $\Theta$ is a separable metric space equipped with the distance $d_\Theta$. Moreover, simple examples show that the infimum in (2.1) might not be attained. It will be often useful to remind that, given any sequence $\{\theta_n\}$

G. Buttazzo et al., *Optimal Urban Networks via Mass Transportation*, Lecture Notes in Mathematics 1961, DOI: 10.1007/978-3-540-85799-0_2,

of paths in $\Theta$ with uniformly bounded Euclidean lengths, by Ascoli–Arzelà Theorem one can find a $\theta \in \Theta$ such that (possibly up to a subsequence) $\theta_n \xrightarrow{d_\Theta} \theta$. This implies, in particular, that the corresponding curves $\theta_n([0,1])$ converge in the Hausdorff distance to $\theta([0,1])$, while the converse implication is not true. Notice that

$$\theta_n \xrightarrow{d_\Theta} \theta \quad \Longrightarrow \quad \mathscr{H}^1\big(\theta([0,1])\big) \leq \liminf_{n\to\infty} \mathscr{H}^1\big(\theta_n([0,1])\big),$$

where $\mathscr{H}^1$ denotes the one-dimensional Hausdorff measure.

In the sequel, for the sake of brevity we will abuse the notation calling $\theta$ also the set $\theta([0,1]) \subseteq \Omega$, when not misleading. We call *endpoints* of the path $\theta$ the points $\theta(0)$ and $\theta(1)$, and, given two paths $\theta_1, \theta_2 \in \Theta$ such that $\theta_1(1) = \theta_2(0)$, the *composition* $\theta_1 \cdot \theta_2$ is defined by the formula

$$\theta_1 \cdot \theta_2(t) := \begin{cases} \theta_1(2t) & \text{for } 0 \leq t \leq 1/2\,, \\ \theta_2(2t-1) & \text{for } 1/2 \leq t \leq 1\,. \end{cases}$$

As already introduced in Chapter 1, we let now $A$, $B : \mathbb{R}^+ \to \mathbb{R}^+$ be the costs of moving by own means and by using the network, i.e. $A(s)$ (resp. $B(s)$) is the cost corresponding to a part of the itinerary of length $s$ covered by own means (resp. with the use of the network). This means that, if the urban network is a Borel set $\Sigma \subseteq \Omega$ of finite length, the total cost of covering a path $\theta \in \Theta$ is given by

$$\delta_\Sigma(\theta) := A\big(\mathscr{H}^1(\theta \setminus \Sigma)\big) + B\big(\mathscr{H}^1(\theta \cap \Sigma)\big)\,, \tag{2.2}$$

since the length $\mathscr{H}^1(\theta\setminus\Sigma)$ is covered by own means and the length $\mathscr{H}^1(\theta\cap\Sigma)$ is covered by the use of the network. Concerning the functions $A$ and $B$, we make from now on the following assumptions:

$$A \text{ is nondecreasing, continuous and } A(0) = 0\,; \tag{2.3}$$

$$B \text{ is nondecreasing, l.s.c. and } B(0) = 0\,. \tag{2.4}$$

Note that these hypotheses follow the intuition: the meaning of the assumptions $A(0) = 0$, $B(0) = 0$ and of the monotonicity are obvious, while the continuity of the function $A$ means that a slightly longer path cannot have a much higher cost, and it is a natural assumption once one moves by own means. On the contrary, a continuity assumption on the function $B$ would rule out some of the most common pricing policies which occur in many real life urban transportation networks: for instance, often such a pricing policy is given by a fixed price (the price of a single ticket) for any positive distance, or is a piecewise constant function.

We define now a "distance" on $\Omega$ which depends on $\Sigma$ and is given by the least cost of the paths connecting two points: in short,

$$d_\Sigma(x,y) := \inf\left\{\delta_\Sigma(\theta) :\ \theta \in \Theta,\, \theta(0) = x,\, \theta(1) = y\right\}. \tag{2.5}$$

The infimum in the above definition is not always attained, as we will see in Example 2.8. Moreover, it has to be pointed out that in general the function $d_\Sigma$ is not a distance; for instance, with $A(s) = B(s) = s^2$ it is easy to see that the triangle inequality does not hold. However, when $A$ and $B$ are subadditive functions, i.e.

$$A(s_1 + s_2) \leq A(s_1) + A(s_2) \text{ for all } s_1,\, s_2 \in \mathbb{R}^+ ,$$
$$B(s_1 + s_2) \leq B(s_1) + B(s_2) \text{ for all } s_1,\, s_2 \in \mathbb{R}^+ ,$$

and they are strictly positive on $(0, +\infty)$, then an easy computation shows that $d_\Sigma$ is in fact a distance (the strict positivity is needed to ensure that $d_\Sigma(x, y) = 0$ implies $x = y$). Nevertheless, with an abuse of notation, we will call $d_\Sigma$ a distance in any case.

**Lemma 2.1.** *For any $\theta \in \Theta$ and any $\varepsilon > 0$, there is a path $\theta_\varepsilon \in \Theta$ such that*

$$\theta_\varepsilon(0) = \theta(0)\,, \qquad \theta_\varepsilon(1) = \theta(1)\,, \qquad d_\Theta(\theta, \theta_\varepsilon) < \varepsilon\,,$$
$$\mathcal{H}^1(\theta_\varepsilon) < \mathcal{H}^1(\theta) + \varepsilon\,, \qquad \mathcal{H}^1(\theta_\varepsilon \cap \Sigma) = 0\,.$$

*Proof.* Since $\Omega \subseteq \mathbb{R}^N$ and $N \geq 2$, we can take a more than countable family $\{\theta_i\}_{i\in I}$ of elements of $\Theta$ such that

- $\theta_i(0) = \theta(0)$ and $\theta_i(1) = \theta(1)$ for each $i \in I$;
- $d_\Theta(\theta, \theta_i) < \varepsilon$ and $\mathcal{H}^1(\theta_\varepsilon) < \mathcal{H}^1(\theta) + \varepsilon$ for each $i \in I$;
- for all $i, j \in I$ with $i \neq j$, $\theta_i \cap \theta_j$ consists of finitely many points.

The proof of this assertion is trivial if the curve $\theta$ is given by a finite union of segments, as Figure 2.1 shows. The general case is now easily achieved approximating any path $\theta$ by a finite union of segments as needed.

The thesis can be then proved making use of the paths $\theta_i$: since $\mathcal{H}^1(\Sigma) < \infty$, the condition $\mathcal{H}^1(\theta_i \cap \Sigma) > 0$ may occur at most for a countable set of indices $i \in I$; one then concludes just by taking one of the remaining paths $\theta_i$. □

**Corollary 2.2.** *For any $\theta \in \Theta$, $\varepsilon > 0$ and $l \leq \mathcal{H}^1(\theta \cap \Sigma)$, there is a path $\theta_{l,\varepsilon} \in \Theta$ such that*

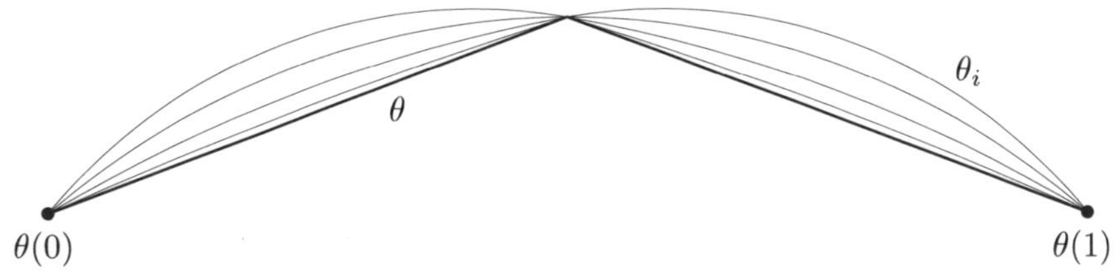


**Fig. 2.1** The path $\theta$ and some paths $\theta_i$

$$\theta_{l,\varepsilon}(0) = \theta(0)\,, \qquad \theta_{l,\varepsilon}(1) = \theta(1)\,, \qquad d_\Theta(\theta, \theta_{l,\varepsilon}) < \varepsilon\,,$$
$$\mathscr{H}^1(\theta_{l,\varepsilon}) < \mathscr{H}^1(\theta) + \varepsilon\,, \qquad \mathscr{H}^1(\theta_{l,\varepsilon} \cap \Sigma) = l\,.$$

*Proof.* This follows easily by Lemma 2.1: let $t \in [0,1]$ be such that

$$\mathscr{H}^1\big(\theta([0,t])\big) = l\,,$$

and define $\theta_1$ to be the restriction of $\theta$ to $[0,t]$ and $\theta_2$ to be the restriction of $\theta$ to $[t,1]$, so that

$$\theta = \theta_1 \cdot \theta_2\,, \qquad \mathscr{H}^1(\theta_1 \cap \Sigma) = l\,.$$

It suffices then to apply Lemma 2.1 to $\theta_2$ and to compose $\theta_1$ with the resulting path. □

**Proposition 2.3.** *The function $d_\Sigma : \Omega \times \Omega \to \mathbb{R}^+$ is continuous.*

*Proof.* This is a consequence of (2.3): take $(x,y) \in \Omega \times \Omega$ and a path $\theta$ between $x$ and $y$ with

$$\delta_\Sigma(\theta) < d_\Sigma(x,y) + \varepsilon\,.$$

Then, given any pair $(\tilde{x}, \tilde{y}) \in \Omega \times \Omega$, we can define a path between $\tilde{x}$ and $\tilde{y}$ by setting $\tilde{\theta} := \alpha \cdot \theta \cdot \beta$ for any choice of paths $\alpha$ and $\beta$ connecting $\tilde{x}$ to $x$ and $y$ to $\tilde{y}$ respectively. Thanks to Lemma 2.1, we may choose $\alpha$ and $\beta$ having $\mathscr{H}^1$-negligible intersection with $\Sigma$ and length less than $|x - \tilde{x}| + \varepsilon$ and $|y - \tilde{y}| + \varepsilon$ respectively. We infer thus

$$\begin{aligned}
d_\Sigma(\tilde{x}, \tilde{y}) \le \delta_\Sigma(\tilde{\theta}) &\le A\big(\mathscr{H}^1(\theta \setminus \Sigma) + |x - \tilde{x}| + |y - \tilde{y}| + 2\varepsilon\big) + B\big(\mathscr{H}^1(\theta \cap \Sigma)\big) \\
&= \delta_\Sigma(\theta) + A\big(\mathscr{H}^1(\theta \setminus \Sigma) + |x - \tilde{x}| + |y - \tilde{y}| + 2\varepsilon\big) - A\big(\mathscr{H}^1(\theta \setminus \Sigma)\big) \\
&\le d_\Sigma(x,y) + \varepsilon + A\big(\mathscr{H}^1(\theta \setminus \Sigma) + |x - \tilde{x}| + |y - \tilde{y}| + 2\varepsilon\big) \\
&\qquad - A\big(\mathscr{H}^1(\theta \setminus \Sigma)\big)\,,
\end{aligned}$$

and the upper semicontinuity of $d_\Sigma$ follows since $\varepsilon > 0$ is arbitrary and $A$ is continuous.

Concerning the lower semicontinuity of $d_\Sigma$, suppose that $x_n \to x$, $y_n \to y$ and that $d_\Sigma(x_n, y_n) \to d$ as $n \to \infty$. This means that there exist paths $\theta_n$ connecting $x_n$ to $y_n$ and satisfying $\delta_\Sigma(\theta_n) \to d$. Composing as before $\theta_n$ with short paths $\alpha_n$ and $\beta_n$ connecting $x$ to $x_n$ and $y_n$ to $y$ respectively, and having

$$\mathscr{H}^1(\alpha_n \cap \Sigma) = \mathscr{H}^1(\beta_n \cap \Sigma) = 0\,,$$

we find the paths $\tilde{\theta}_n$ between $x$ and $y$ satisfying

$$\delta_\Sigma(\tilde{\theta}_n) = \delta_\Sigma(\theta_n) + A\big(\mathscr{H}^1(\theta_n \setminus \Sigma) + \mathscr{H}^1(\alpha_n) + \mathscr{H}^1(\beta_n)\big) - A\big(\mathscr{H}^1(\theta_n \setminus \Sigma)\big)\,.$$

Since $\delta_\Sigma(\theta_n) \to d$ and since

$$\mathscr{H}^1(\alpha_n) + \mathscr{H}^1(\beta_n) \to 0\,,$$

the conclusion follows if $\mathscr{H}^1(\theta_n \backslash \Sigma)$ is uniformly bounded, because $A$ is continuous hence uniformly continuous on compact sets. At last, if $\mathscr{H}^1(\theta_n \setminus \Sigma)$ is not uniformly bounded, then

$$\mathscr{H}^1(\theta_n \setminus \Sigma) > |x-y| + 1$$

for $n$ arbitrarily large; in this case, we could directly take a path $\theta$ close to the segment connecting $x$ to $y$ and having negligible intersection with $\Sigma$, so that

$$\delta_\Sigma(\theta) = A\big(\mathscr{H}^1(\theta)\big) \le A\big(|x-y|+1\big) \le A\big(\mathscr{H}^1(\theta_n \setminus \Sigma)\big) \le \delta_\Sigma(\theta_n)\,,$$

and hence, the thesis follows in this case too. □

The problem we want to study is to find the best transportation network $\Sigma$ to move the population from their "homes" to their "workplaces". To set the problem, we consider two probability measures $f^+$, $f^-$ on $\Omega$ describing the distributions of homes and workplaces respectively. The following notion is often used in transportation theory; throughout the monograph, $\pi_i : \Omega \times \Omega \to \Omega$, $i = 1, 2$, stands for the $i$–th projection, and for a Borel map $g : X \to Y$ the push-forward $g_\# : \mathcal{M}^+(X) \to \mathcal{M}^+(Y)$ is defined by

$$g_\#\mu(A) := \mu\big(g^{-1}(A)\big) \qquad\qquad \text{for any Borel set } A \subseteq Y\,,$$

where $\mathcal{M}^+(Z)$ is the space of the finite positive measures on a generic space $Z$ (see Appendix B.1).

**Definition 2.4.** A *transport plan* is a positive measure $\gamma \in \mathcal{M}^+(\Omega \times \Omega)$, the marginals of which are $f^+$ and $f^-$, i.e.

$$\pi_{1\#}\gamma = f^+\,, \qquad\qquad \pi_{2\#}\gamma = f^-\,.$$

One can intuitively think that $\gamma(x, y)$ is the number of people moving from $x$ to $y$, or, more precisely, that $\gamma(C \times D)$ is the number of people living in $C \subseteq \Omega$ and working in $D \subseteq \Omega$. To each transport plan $\gamma$ we associate the total cost of transportation according to the formula

$$I_\Sigma(\gamma) := \iint_{\Omega\times\Omega} d_\Sigma(x, y)\, d\gamma(x, y)\,. \tag{2.6}$$

The Monge-Kantorovich optimal transport problem consists in finding a transport plan $\bar\gamma \in \mathcal{M}^+(\Omega \times \Omega)$ (which is usually called *optimal transport plan*) minimizing $I_\Sigma$.

It is important to notice that the transport plan $\gamma$ gives no precise information on *how* the mass is moving (i.e. which trajectories are chosen for transportation). To be able to recover such an information we will make use of the following definition, already used in [58] (a quite similar idea was already used elsewhere, for instance in [24, 65, 6]).

**Definition 2.5.** A *transport path measure* (shortly t.p.m. in the sequel) is a measure $\eta \in \mathcal{M}^+(\Theta)$ with the property that its first and last projections are $f^+$ and $f^-$, i.e.

$$p_{0\#}\eta = f^+ \qquad\qquad p_{1\#}\eta = f^-, \tag{2.7}$$

where for $t = 0, 1$ we denote by $p_t : \Theta \to \Omega$ the function $p_t(\theta) := \theta(t)$.

It is important to understand the meaning of the above definition: roughly speaking, if $\eta$ is a t.p.m., then $\eta(\theta)$ indicates the amount of mass to be moved along the path $\theta$; more precisely, $\eta(E)$ is the mass following the paths contained in $E \subseteq \Theta$. The meaning of condition (2.7) is then clear, since $p_{0\#}\eta$ and $p_{1\#}\eta$ are respectively the measure from which $\eta$ starts and the measure to which it is transported.

We are now able to define the total cost of transportation associated to any t.p.m. by the formula

$$C_\Sigma(\eta) := \int_\Theta \delta_\Sigma(\theta)\, d\eta(\theta). \tag{2.8}$$

Finally, we denote by $MK(\Sigma)$ the infimum of the above costs, namely,

$$MK(\Sigma) := \inf \{C_\Sigma(\eta) \ : \ \eta \text{ is a t.p.m.}\}. \tag{2.9}$$

The purpose of this monograph is to study the problem of finding the best possible network $\Sigma$: in other words, we want to find a set $\Sigma$ having minimal total cost of usage (defined below). To do that, as already discussed in Chapter 1, we consider a function $H : \mathbb{R}^+ \to \overline{R}^+$, where $H(l)$ represents the maintenance cost of a network $\Sigma$ of length $\mathscr{H}^1(\Sigma) = l$. We assume on $H$ the natural conditions

$$H \text{ is nondecreasing, l.s.c., } H(0) = 0 \text{ and } H(l) \to \infty \text{ as } l \to \infty. \tag{2.10}$$

Finally, the *total cost* of usage of $\Sigma$ is defined by the formula

$$\mathfrak{F}(\Sigma) := MK(\Sigma) + H\big(\mathscr{H}^1(\Sigma)\big). \tag{2.11}$$

Our goal is to study the problem of minimizing the functional $\mathfrak{F}$.

## 2.2 Properties of Optimal Paths and Relaxed Costs

In (2.5) we defined a distance in $\Omega$ as the infimum of the costs of the paths connecting two given points. We show now the possibility to choose a Borel selection of paths which have almost minimal costs in the sense of proposition below.

**Proposition 2.6.** *For any $\varepsilon > 0$ there is a Borel function $q_\varepsilon : \Omega \times \Omega \to \Theta$ such that $q_\varepsilon(x, y)$ is a path connecting $x$ to $y$ with*

$$\delta_\Sigma\big(q_\varepsilon(x, y)\big) < d_\Sigma(x, y) + \varepsilon\,. \tag{2.12}$$

*Proof.* Fix a $\rho > 0$ and let $\{x_i\}$ be a finite set of points in $\Omega$ such that

$$\bigcup B(x_i, \rho) \supseteq \Omega\,.$$

Let then $C_{ij} \subseteq \Omega \times \Omega$ be pairwise disjoint Borel sets covering $\Omega \times \Omega$, each contained in $B\big((x_i, x_j), 2\rho\big)$. Now, given $i$, $j$, let $\theta_{ij} \in \Theta$ be a path connecting $x_i$ to $x_j$ and having a cost minimal up to an error $\rho$, that is

$$\delta_\Sigma(\theta_{ij}) < d_\Sigma(x_i, x_j) + \rho\,.$$

We claim that the conclusion follows if for every $x \in \Omega$ there is a Borel map

$$\alpha_x : B(x, 2\rho) \to \Theta$$

such that $\alpha_x(y)$ is a path between $x$ and $y$ with length less than $4\rho$ and having $\mathscr{H}^1$−negligible intersection with $\Sigma$. Indeed, defining on each $C_{ij}$ the function $q_\varepsilon$ by the formula

$$q_\varepsilon(x, y) := \widehat{\alpha_{x_i}(x)} \cdot \theta_{ij} \cdot \alpha_{x_j}(y)$$

(where $\hat{\theta}(t) := \theta(1 - t)$), one has that $q_\varepsilon$ is a Borel function; moreover, if $\rho$ is sufficiently small, one gets (2.12) by the continuity of $A$. It suffices therefore to prove the existence of such an $\alpha_x$ (observe that Lemma 2.1 already provides a map satisfying all the required conditions except for the Borel property). For this purpose, we begin defining $\alpha_x(y)$ as the line segment between $x$ and $y$. Since $\Sigma$ is rectifiable, such a segment has $\mathscr{H}^1$-negligible intersection with $\Sigma$ unless $y$ is contained in one of countably many radii $\{R_k\}_{k\in\mathbb{N}}$ of the ball $B(x, 2\rho)$. For each $k \in \mathbb{N}$, choose arbitrarily a two-dimensional halfplane $\Pi_k$ containing $R_k$ on its boundary; then, for $y \in R_k$, define $\alpha_x(y)$ as the half circle joining $x$ to $y$ and lying on $\Pi_k$. Arguing as before, it is clear that such a path has $\mathscr{H}^1$-negligible intersection with $\Sigma$ except for countably many points $y \in R_k$. Finally, for each of these latter $y$, by Lemma 2.1 we may arbitrarily select a path $\alpha_x(y)$ connecting $x$ to $y$ which is shorter than $4\rho$ and has $\mathscr{H}^1$–negligible intersection with $\Sigma$. The resulting function $\alpha_x$ has the required properties and so the proof is completed. □

**Corollary 2.7.** *For any $\varepsilon > 0$ there is a Borel function $q'_\varepsilon : \Omega \times \Omega \to \Theta$ such that $q'_\varepsilon(x, y)$ is a path connecting $x$ with $y$ and satisfying*

$$\mathscr{H}^1(q'_\varepsilon(x, y)) \le |y - x| + \varepsilon\,, \qquad \mathscr{H}^1(q'_\varepsilon(x, y) \cap \Sigma) = 0\,.$$

*Proof.* Consider the case when

$$A(s) = s\,, \qquad B(s) = \operatorname{diam} \Omega + 2\varepsilon$$

for every $s > 0$. By Lemma 2.1 it is clear that $d_\Sigma(x, y) = |y - x|$ and that $\delta_\Sigma(\theta) = \mathscr{H}^1(\theta)$ whenever $\mathscr{H}^1(\theta \cap \Sigma) = 0$. Apply now Proposition 2.6 to find a map $q'_\varepsilon$ such that

$$\delta_\Sigma(q'_\varepsilon(x, y)) < d_\Sigma(x, y) + \varepsilon = |y - x| + \varepsilon\,.$$

If

$$\mathscr{H}^1(q'_\varepsilon(x, y) \cap \Sigma) > 0\,,$$

then

$$\delta_\Sigma(q'_\varepsilon(x, y)) \ge \operatorname{diam} \Omega + 2\varepsilon > |y - x| + \varepsilon\,,$$

and this gives a contradiction. Thus,

$$\mathscr{H}^1(q'_\varepsilon(x, y) \cap \Sigma) = 0$$

and, as a consequence,

$$\mathscr{H}^1(q'_\varepsilon(x, y)) = \delta_\Sigma(q'_\varepsilon(x, y)) < |y - x| + \varepsilon\,;$$

hence the thesis follows. □

We see now an example, showing that the infimum in (2.5) may be not a minimum, and that $\delta_\Sigma$ may be not lower semicontinuous.

*Example 2.8.* Let $\Omega$ be the ball in $\mathbb{R}^2$ centered at the origin and with radius 2, let $\Sigma = [0, 1] \times \{0\}$, $A(t) = t$ and $B(t) = 2t$; let moreover $\theta$ and $\theta_n$ be the paths connecting $(0, 0)$ to $(1, 0)$ given by

$$\theta(t) := (t, 0)\,, \qquad \theta_n(t) := \left(t, \frac{1 - |2t - 1|}{n}\right).$$

Then one has that $\theta_n$ converges to $\theta$ in $(\Theta, d_\Theta)$, $\delta_\Sigma(\theta) = 2$, while $\delta_\Sigma(\theta_n) \to 1$: therefore, $\delta_\Sigma$ is not l.s.c. Moreover, it is clear that

$$d_\Sigma\big((0, 0), (1, 0)\big) = 1\,,$$

but $\delta_\Sigma(\sigma) > 1$ for each path $\sigma \in \Theta$ connecting $(0,0)$ and $(1,0)$. Hence, the infimum in (2.5) is not a minimum.

Since $\delta_\Sigma$ is not, in general, l.s.c., we compute now its relaxed envelope with fixed endpoints,

$$\bar{\delta}_\Sigma(\theta) := \inf \left\{ \liminf_{n\to\infty} \delta_\Sigma(\theta_n) : \theta_n(0) = \theta(0),\, \theta_n(1) = \theta(1),\, \theta_n \stackrel{\Theta}{\to} \theta \right\} . \quad (2.13)$$

Notice that $\bar{\delta}_\Sigma \leq \delta_\Sigma$, and that the infimum in (2.13) is a minimum. Thanks to the standard properties of relaxed envelopes (see [16]), we are allowed to rewrite (2.5) obtaining

$$d_\Sigma(x,y) = \inf \left\{ \bar{\delta}_\Sigma(\theta) : \ \theta \in \Theta,\, \theta(0) = x,\, \theta(1) = y \right\} . \quad (2.14)$$

**Proposition 2.9.** *The function $\bar{\delta}_\Sigma : \Theta \to \mathbb{R}^+$ is l.s.c.*

*Proof.* Let us take $\theta_n \to \theta$ in $\Theta$: then, without loss of generality, we may assume

$$|\theta_n(0) - \theta(0)| \leq \frac{1}{n}, \qquad |\theta_n(1) - \theta(1)| \leq \frac{1}{n}.$$

Following (2.13), we choose $\hat{\theta}_n$ having the same endpoints as $\theta_n$ and such that

$$d_\Theta(\theta_n, \hat{\theta}_n) \leq \frac{1}{n}, \qquad \delta_\Sigma(\hat{\theta}_n) \leq \bar{\delta}_\Sigma(\theta_n) + \frac{1}{n}. \quad (2.15)$$

Take now, according to Lemma 2.1, two paths $\alpha_n$ and $\beta_n$ connecting $\theta(0)$ with $\hat{\theta}_n(0)$ and $\hat{\theta}_n(1)$ with $\theta(1)$ respectively, with the properties

$$\begin{aligned} &\mathcal{H}^1(\alpha_n \setminus \Sigma) \leq \frac{2}{n}, \qquad &&\mathcal{H}^1(\beta_n \setminus \Sigma) \leq \frac{2}{n}, \\ &\mathcal{H}^1(\alpha_n \cap \Sigma) = 0, &&\mathcal{H}^1(\beta_n \cap \Sigma) = 0. \end{aligned} \quad (2.16)$$

Define then $\overline{\theta}_n := \alpha_n \cdot \hat{\theta}_n \cdot \beta_n$, so that $\{\overline{\theta}_n\}_{n\in\mathbb{N}}$ is a sequence of paths connecting $\theta(0)$ to $\theta(1)$ which still converges to $\theta$. For any $n \in \mathbb{N}$, by (2.15) and (2.16) we have

$$\begin{aligned} \delta_\Sigma(\overline{\theta}_n) &= A\Big(\mathcal{H}^1(\overline{\theta}_n \setminus \Sigma)\Big) + B\Big(\mathcal{H}^1(\overline{\theta}_n \cap \Sigma)\Big) \\ &= A\Big(\mathcal{H}^1((\hat{\theta}_n \cup \alpha_n \cup \beta_n) \setminus \Sigma)\Big) + B\Big(\mathcal{H}^1(\hat{\theta}_n \cap \Sigma)\Big) \\ &\leq \delta_\Sigma(\hat{\theta}_n) + A\Big(\mathcal{H}^1(\hat{\theta}_n \setminus \Sigma) + 4/n\Big) - A\Big(\mathcal{H}^1(\hat{\theta}_n \setminus \Sigma)\Big) \\ &\leq \bar{\delta}_\Sigma(\theta_n) + 1/n + A\Big(\mathcal{H}^1(\hat{\theta}_n \setminus \Sigma) + 4/n\Big) - A\Big(\mathcal{H}^1(\hat{\theta}_n \setminus \Sigma)\Big). \end{aligned}$$

Since the paths $\{\theta_n\}$ have uniformly bounded lengths, by the uniform continuity of $A$ in the bounded intervals and by (2.13) we infer

$$\bar{\delta}_\Sigma(\theta) \leq \liminf_{n\to\infty} \delta_\Sigma(\overline{\theta}_n) \leq \liminf_{n\to\infty} \bar{\delta}_\Sigma(\theta_n),$$

so the proof is completed. □

**Corollary 2.10.** *The l.s.c. envelope of $\delta_\Sigma$ in $(\Theta, d_\Theta)$ is $\bar{\delta}_\Sigma$.*

*Proof.* The l.s.c. envelope of $\delta_\Sigma$ in $(\Theta, d_\Theta)$ is lower than $\bar{\delta}_\Sigma$, as a direct consequence of the definition (2.13). On the other hand, it is the greatest l.s.c. function lower than $\delta_\Sigma$, thus it is also greater than $\bar{\delta}_\Sigma$ by Proposition 2.9. □

**Corollary 2.11.** *The infimum in (2.14) is actually a minimum.*

*Proof.* Let us choose $x$ and $y$ and take a minimizing sequence $\theta_n$ for (2.14): if the Euclidean lengths of $\theta_n$ (possibly up to a subsequence) are bounded, then the result immediately follows from the lower semicontinuity of $\bar{\delta}_\Sigma$ and by Ascoli–Arzelà Theorem. Otherwise, since $\Sigma$ has finite length, it would follow that

$$\limsup \mathscr{H}^1(\theta_n \setminus \Sigma) = \infty;$$

in this case, take a path $\theta$ joining $x$ to $y$ with $\mathscr{H}^1$–negligible intersection with $\Sigma$ and with finite length: since $A$ is nondecreasing, this path provides the minimum in (2.14). □

More precisely, we see that one can somehow "pass to the limit" in Proposition 2.6. Throughout the monograph, we will call *geodesics* the paths $\theta$ such that

$$\bar{\delta}_\Sigma(\theta) = d_\Sigma\big(\theta(0), \theta(1)\big).$$

**Corollary 2.12.** *There is a Borel function $q : \Omega \times \Omega \to \Theta$ such that $q(x,y)$ is a path connecting $x$ to $y$ with cost $\bar{\delta}_\Sigma\big(q(x,y)\big) = d_\Sigma(x,y)$.*

*Proof.* Using the classical results in [28], it is sufficient to show that the subset $G$ of $\Theta$ given by the geodesics is closed and there is at least one element of $G$ connecting any couple of points in $\Omega \times \Omega$. The second fact follows from Corollary 2.11, while the closedness of $G$ is a direct consequence of the lower semicontinuity of $\bar{\delta}_\Sigma$ and of the continuity of $d_\Sigma$. □

**Lemma 2.13.** *For any $\varepsilon > 0$, there is a Borel function $\alpha_\varepsilon : \Theta \to \Theta$ such that for any $\theta \in \Theta$ one has*

$$\begin{aligned}
&\big(\alpha_\varepsilon(\theta)\big)(0) = \theta(0), && \big(\alpha_\varepsilon(\theta)\big)(1) = \theta(1), && d_\Theta(\alpha_\varepsilon(\theta), \theta) \leq \varepsilon, \\
&\mathscr{H}^1(\alpha_\varepsilon(\theta)) \leq \mathscr{H}^1(\theta) + \varepsilon, && \delta_\Sigma(\alpha_\varepsilon(\theta)) \leq \bar{\delta}_\Sigma(\theta) + \varepsilon.
\end{aligned}$$

*Proof.* Our argument is quite similar to the one in Proposition 2.6: fixed $L > 0$ and fixed arbitrarily a path $\theta \in \Theta$ with $\mathscr{H}^1(\theta) \leq L$, we know by definition of $\bar{\delta}_\Sigma$ the existence of a path $\tilde{\theta}$ with

$$\begin{aligned} &\tilde{\theta}(0) = \theta(0)\,, && \tilde{\theta}(1) = \theta(1)\,, && d_\Theta(\tilde{\theta}, \theta) \leq \frac{\varepsilon}{4}\,, \\ &\mathscr{H}^1(\tilde{\theta}) \leq \mathscr{H}^1(\theta) + \frac{\varepsilon}{4}\,, && \delta_\Sigma(\tilde{\theta}) \leq \bar{\delta}_\Sigma(\theta) + \frac{\varepsilon}{4}\,. \end{aligned} \tag{2.17}$$

We take now a number $\delta \leq \varepsilon/8$ such that

$$A(s + 4\delta) - A(s) \leq \frac{\varepsilon}{2}$$

for any $0 \leq s \leq L$, which is possible by the continuity of $A$; moreover, since the Euclidean length and the map $\bar{\delta}_\Sigma$ are l.s.c., we can also assume that $\delta$ is so small that

$$\begin{cases} \mathscr{H}^1(\sigma) \geq \mathscr{H}^1(\theta) - \dfrac{\varepsilon}{4}\,, \\ \bar{\delta}_\Sigma(\sigma) \geq \bar{\delta}_\Sigma(\theta) - \dfrac{\varepsilon}{4}\,, \end{cases} \qquad \text{whenever } d_\Theta(\theta, \sigma) \leq \delta\,. \tag{2.18}$$

If we can define a Borel function $\alpha_\varepsilon : B_\Theta(\theta, \delta) \to \Theta$ as in the claim of this corollary, this will show the thesis: indeed, since the subset $\Theta_L$ of $\Theta$ made by the paths of Euclidean length bounded by $L$ is compact, it can be covered by a finite number of balls $B_\Theta(\theta_i, \delta_i)$, so that we infer the existence of a Borel function $\alpha_\varepsilon : \Theta_L \to \Theta$ as in the claim; finally, it is immediate to conclude the thesis covering $\Theta$ with countably many sets $\Theta_{L_i}$ for a sequence $L_i \to \infty$. Summarizing, we can restrict our attention to a ball $B_\Theta(\theta, \delta)$.

Define now the Borel function $\beta_1 : B_\Theta(\theta, \delta) \to \Theta$ as

$$\beta_1(\sigma) := q'_\delta\big(\sigma(0), \theta(0)\big)\,,$$

where $q'_\delta$ is as in Corollary 2.7: then $\beta_1(\sigma)$ is a path connecting $\sigma(0)$ with $\theta(0)$ such that

$$\mathscr{H}^1(\beta_1(\sigma) \cap \Sigma) = 0\,, \qquad \mathscr{H}^1(\beta_1(\sigma)) \leq |\sigma(0) - \theta(0)| + \delta \leq 2\delta\,. \tag{2.19}$$

Similarly, we let $\beta_2 : B_\Theta(\theta, \delta) \to \Theta$ to be a Borel function such that $\beta_2(\sigma)$ is a path connecting $\theta(1)$ with $\sigma(1)$ satisfying

$$\mathscr{H}^1(\beta_2(\sigma) \cap \Sigma) = 0\,, \qquad \mathscr{H}^1(\beta_2(\sigma)) \leq 2\delta\,; \tag{2.20}$$

We finally define $\alpha_\varepsilon(\sigma) := \beta_1(\sigma) \cdot \tilde{\theta} \cdot \beta_2(\sigma)$: by construction, the map

$$B_\Theta(\theta, \delta) \ni \sigma \mapsto \alpha_\varepsilon(\sigma) \in \Theta$$

is Borel; moreover,

$$\alpha_\varepsilon(\sigma(0)) = \sigma(0)\,, \qquad\qquad \alpha_\varepsilon(\sigma(1)) = \sigma(1)\,.$$

In addition, minding (2.19), (2.20) and (2.17), we get

$$d_\Theta(\alpha_\varepsilon(\sigma),\sigma) \le d_\Theta(\alpha_\varepsilon(\sigma),\tilde\theta) + d_\Theta(\tilde\theta,\theta) + d_\Theta(\theta,\sigma) \le 2\delta + \frac{\varepsilon}{4} + \delta < \varepsilon\,.$$

Again by (2.19), (2.20), (2.17) and (2.18) one has

$$\mathscr{H}^1(\alpha_\varepsilon(\sigma)) \le 4\delta + \mathscr{H}^1(\tilde\theta) \le 4\delta + \frac{\varepsilon}{4} + \mathscr{H}^1(\theta) \le \mathscr{H}^1(\sigma) + \varepsilon\,.$$

Finally, by (2.19) and (2.20) we know that

$$\mathscr{H}^1(\alpha_\varepsilon(\sigma)\cap\Sigma) = \mathscr{H}^1(\tilde\theta\cap\Sigma)$$

so that again (2.19) and (2.20), together with (2.17) and (2.18), yield

$$\begin{aligned}\delta_\Sigma(\alpha_\varepsilon(\sigma)) &\le A\big(\mathscr{H}^1(\tilde\theta\setminus\Sigma)+4\delta\big) + B\big(\mathscr{H}^1(\tilde\theta\cap\Sigma)\big)\\ &\le \delta_\Sigma(\tilde\theta) + A\big(\mathscr{H}^1(\tilde\theta\setminus\Sigma)+4\delta\big) - A\big(\mathscr{H}^1(\tilde\theta\setminus\Sigma)\big)\\ &\le \delta_\Sigma(\tilde\theta) + \frac{\varepsilon}{2} \le \bar\delta_\Sigma(\theta) + \frac{3}{4}\varepsilon \le \bar\delta_\Sigma(\sigma)+\varepsilon\,:\end{aligned}$$

hence, the proof is complete. □

Now, generalizing (2.8), set

$$\overline{C}_\Sigma(\eta) := \int_\Theta \bar\delta_\Sigma(\theta)\,d\eta(\theta)\,. \tag{2.21}$$

**Proposition 2.14.** *The following equalities hold*

$$\begin{aligned}\inf\big\{C_\Sigma(\eta):\ \eta \text{ is a t.p.m.}\big\} &= \min\{I_\Sigma(\gamma)\ :\ \gamma \text{ is a transport plan}\}\\ &= \min\big\{\overline{C}_\Sigma(\eta):\ \eta \text{ is a t.p.m.}\big\} \qquad \big(=MK(\Sigma)\big)\,.\end{aligned} \tag{2.22}$$

Before giving the proof, we point out the following remark.

*Remark 2.15.* The equality (2.22) ensures the existence of at least one optimal transport plan $\gamma_{\text{opt}}$ and one t.p.m. $\eta_{\text{opt}}$ optimal with respect to $\overline{C}_\Sigma$, which satisfy the equality $I_\Sigma(\gamma_{\text{opt}}) = \overline{C}_\Sigma(\eta_{\text{opt}})$. On the other hand, the infimum in (2.22) needs not to be achieved: for instance, just consider the situation of Example 2.8 with $f^+ := \delta_{(0,0)}$ and $f^- := \delta_{(1,0)}$.

Concerning the equality between the two minima in (2.22), in particular, if $\gamma_{\text{opt}}$ is an optimal transport plan then $q_\#\gamma_{\text{opt}}$ is an optimal t.p.m., where $q$ is defined in Corollary 2.12. Conversely, if $\eta_{\text{opt}}$ is an optimal t.p.m. then $(p_0,p_1)_\#\eta_{\text{opt}}$ is an optimal transport plan, where $p_0$ and $p_1$ are as in Definition 2.5.

*Proof (of Proposition 2.14).* First of all, note that the set of all transport plans is a bounded and weakly* closed subset of $\mathcal{M}^+(\Omega \times \Omega)$; hence, it is weakly* compact by tightness (see Appendix B.1). Moreover, $I_\Sigma$ is a continuous function on $\mathcal{M}(\Omega \times \Omega)$ with respect to the weak* topology thanks to Proposition 2.3. Therefore, the existence of some optimal transport plan is straightforward.

Given now a t.p.m. $\eta$, one can construct the associated transport plan $\gamma = (p_0, p_1)_\# \eta$, and from (2.5) we get $I_\Sigma(\gamma) \le C_\Sigma(\eta)$. On the other hand, given any transport plan $\gamma$ and $\varepsilon > 0$, we can define $\eta := q_{\varepsilon\#}\gamma$ where $q_\varepsilon$ is as in Proposition 2.6; we obtain $C_\Sigma(\eta) \le I_\Sigma(\gamma) + \varepsilon$, thus the first equality in (2.22) is established.

Concerning the second one, using (2.14) in place of (2.5) in the previous argument one gets

$$\min\{I_\Sigma(\gamma)\} \le \inf\{\overline{C}_\Sigma(\eta)\}\,.$$

But since $\overline{C}_\Sigma \le C_\Sigma$ (because $\bar{\delta}_\Sigma \le \delta_\Sigma$), it is also true that

$$\inf\{\overline{C}_\Sigma(\eta)\} \le \inf\{C_\Sigma(\eta)\}\,.$$

We derive $\min\{I_\Sigma(\gamma)\} = \inf\{\overline{C}_\Sigma(\eta)\}$, so to conclude we need only to prove that the last inf is a minimum. To this aim, it suffices to take an optimal transport plan $\gamma_{\mathrm{opt}}$ and to define $\eta := q_\# \gamma_{\mathrm{opt}}$ where $q$ is as in Corollary 2.12: by definition of $q$, one has $\overline{C}_\Sigma(\eta) = I_\Sigma(\gamma_{\mathrm{opt}})$, so $\eta$ minimizes $\overline{C}_\Sigma$ and the proof is achieved. □

From now on we will often say that a set $\Delta \subseteq \Theta$ is *bounded in* $\Theta$ *by* $L$ if for any $\theta \in \Delta$ we have $\mathscr{H}^1(\theta) \le L$; we will also say that $\Delta$ is a *bounded subset of* $\Theta$ if it is bounded in $\Theta$ by some constant $L$. Notice that this last definition does not coincide with the usual boundedness in $\Theta$ with respect to the distance $d_\Theta$, which we will never consider; in fact, this last notion of boundedness would be useless, since the whole set $\Theta$ is clearly bounded with respect to $d_\Theta$ by the diameter of $\Omega$. We recall that, as already mentioned at the beginning of Section 2.1, the bounded subsets of $\Theta$ are sequentially compact with respect to $d_\Theta$; this becomes particularly helpful once we know that a t.p.m. is concentrated on a bounded subset of $\Theta$, which is the argument of Corollary 2.17 below.

**Lemma 2.16.** *If $A(s)$ is not constant for large $s$ (for instance, if $A(s) \to \infty$ as $s \to \infty$), then there is a constant $L \in \mathbb{R}$ such that the Euclidean length $\mathscr{H}^1(\theta)$ of any geodesic $\theta$ is bounded by $L$. Otherwise, if $A(s)$ is constant for large $s$, it is still true that for any pair $(x, y)$ of points in $\Omega$ there exists some geodesic of length bounded by $L$. In both cases, the constant $L$ depends only on $A$, $\Omega$ and $\mathscr{H}^1(\Sigma)$ (but not on $\Sigma$).*

*Proof.* Suppose first that $A(s)$ is not constant for large $s$, and let $L$ be a sufficiently large number such that

$$A\big(L - \mathscr{H}^1(\Sigma)\big) > A\big(\operatorname{diam}\Omega + 1\big)\,.$$

Take now a path $\theta \in \Theta$ with $\mathscr{H}^1(\theta) > L$ and let $\hat{\theta}$, according to Lemma 2.1, be a path with length less than

$$|\theta(1) - \theta(0)| + 1 \leq \operatorname{diam} \Omega + 1$$

connecting $\theta(0)$ to $\theta(1)$ and having $\mathscr{H}^1$–negligible intersection with $\Sigma$. Since

$$\mathscr{H}^1(\theta \setminus \Sigma) \geq \mathscr{H}^1(\theta) - \mathscr{H}^1(\Sigma) > L - \mathscr{H}^1(\Sigma),$$

we immediately get $\bar{\delta}_\Sigma(\theta) > \bar{\delta}_\Sigma(\hat{\theta})$, so that $\theta$ is not a geodesic and the first part of the proof is achieved.

Consider now the case when $A(s)$ is constant for large $s$, and let

$$L := \mathscr{H}^1(\Sigma) + \operatorname{diam} \Omega + 1 .$$

Arguing exactly as in the first part of the proof, we see that for any path $\theta$ there is a path $\hat{\theta}$ with $\mathscr{H}^1(\hat{\theta}) \leq \operatorname{diam} \Omega + 1$ and with $\bar{\delta}_\Sigma(\hat{\theta}) \leq \bar{\delta}_\Sigma(\theta)$ (the only difference is that this time the strict inequality $\bar{\delta}_\Sigma(\hat{\theta}) < \bar{\delta}_\Sigma(\theta)$ in the case $\mathscr{H}^1(\theta) > L$ may be false). Hence, it is not true that *all* the geodesics have Euclidean length less than $L$, but that for any pair $(x, y) \in \Omega \times \Omega$ there is *at least one* geodesic between $x$ and $y$ of Euclidean length less than $L$. □

**Corollary 2.17.** *If $A(s)$ is not constant for large $s$ then the support of any t.p.m. $\eta$ which is optimal with respect to $\overline{C}_\Sigma$ is bounded in $\Theta$ by $L$, where $L$ is as in the previous Lemma. Otherwise, if $A(s)$ is constant for large $s$, it is still true that there exists some optimal t.p.m. $\eta$ the support of which is bounded in $\Theta$ by $L$.*

*Proof.* Recall that, thanks to (2.22), any t.p.m. optimal with respect to $\overline{C}_\Sigma$ is concentrated in the set of all geodesics; this set is closed, as already noticed in Corollary 2.12, hence the whole support of any optimal t.p.m. is made by geodesics and the first part of the proof is trivial.

Concerning the second claim, we recall that Corollary 2.12 implies that the set $G$ of all geodesics is a closed subset of $\Theta$ containing at least one path which connects any given pair of points in $\Omega \times \Omega$. The same property is true for the set

$$G_L := G \cap \{\theta \in \Theta : \mathscr{H}^1(\theta) \leq L\},$$

by the above lemma and since the Euclidean length is l.s.c. with respect to the distance in $\Theta$. Therefore, arguing as in Corollary 2.12, we find a Borel map $\tilde{q} : \Omega \times \Omega \to \Theta$ such that $\tilde{q}(x, y)$ is a geodesic between $x$ and $y$ of Euclidean length less than $L$. This easily gives also the second part of the thesis: arguing as in Proposition 2.14, taken any optimal transport plan $\gamma$, one has that the t.p.m. $\tilde{q}_\# \gamma$ is as required. □

We present now a useful exact formula for $\bar{\delta}_\Sigma$.

**Proposition 2.18.** *The following equality holds:*

$$\bar{\delta}_\Sigma(\theta) = J\big(\mathscr{H}^1(\theta \setminus \Sigma), \mathscr{H}^1(\theta \cap \Sigma)\big)\,, \tag{2.23}$$

*where the function* $J\colon \mathbb{R}^+ \times \mathbb{R}^+ \to \mathbb{R}$ *is given by*

$$J(a,b) := \inf \big\{ A\big(a+l\big) + B\big(b-l\big) :\ 0 \le l \le b \big\}\,. \tag{2.24}$$

Before giving the proof, we shortly discuss the above formula.

*Remark 2.19.* The meaning of (2.23), as one can understand comparing with (2.2), is that, roughly speaking, one can "walk on the railway": in other words, the cost $\bar{\delta}_\Sigma$ of some path $\theta$ is not necessarily given by the cost of moving by own means out of the network and by train along it, but moving by own means out of the network and possibly in some part of it, and by train along the remaining part. The basic idea of the proof is then easily imagined: instead of walking on the network, one can just walk very close to it, which is possible since the dimension $N$ is larger than 1.

*Proof (of Proposition 2.18).* Set $a := \mathscr{H}^1(\theta \setminus \Sigma)$ and $b := \mathscr{H}^1(\theta \cap \Sigma)$, then take an arbitrary sequence $\theta_n$ of paths having the same endpoints as $\theta$ and converging to $\theta$. It is known that

$$\mathscr{H}^1(\theta) \le \liminf_{n\to\infty} \mathscr{H}^1(\theta_n)\,, \tag{2.25}$$

$$\mathscr{H}^1(\theta \setminus \Sigma) \le \liminf_{n\to\infty} \mathscr{H}^1(\theta_n \setminus \Sigma)\,; \tag{2.26}$$

the first inequality is the classical lower semicontinuity of the length, the second is a recent generalization of the Gołąb theorem that we state in Theorem 3.6 (see also for instance [14] and [30]).

For a given $n \in \mathbb{N}$, assume that

$$\mathscr{H}^1(\theta_n \cap \Sigma) \ge \mathscr{H}^1(\theta \cap \Sigma)\ :$$

then, taking $l = 0$ in (2.24), we obtain

$$\begin{aligned} J(a,b) &\le A\big(\mathscr{H}^1(\theta \setminus \Sigma)\big) + B\big(\mathscr{H}^1(\theta \cap \Sigma)\big) \\ &\le A\big(\mathscr{H}^1(\theta \setminus \Sigma)\big) + B\big(\mathscr{H}^1(\theta_n \cap \Sigma)\big) \\ &\le \delta_\Sigma(\theta_n) + A\big(\mathscr{H}^1(\theta \setminus \Sigma)\big) - A\big(\mathscr{H}^1(\theta_n \setminus \Sigma)\big)\,. \end{aligned} \tag{2.27}$$

On the other hand, if

$$\mathscr{H}^1(\theta_n \cap \Sigma) < \mathscr{H}^1(\theta \cap \Sigma)$$

then, taking

$$l := \mathscr{H}^1(\theta \cap \Sigma) - \mathscr{H}^1(\theta_n \cap \Sigma)$$

in (2.24), we obtain

$$\begin{aligned} J(a,b) &\le A\big(\mathscr{H}^1(\theta) - \mathscr{H}^1(\theta_n \cap \Sigma)\big) + B\big(\mathscr{H}^1(\theta_n \cap \Sigma)\big) \\ &\le \delta_\Sigma(\theta_n) + A\big(\mathscr{H}^1(\theta) - \mathscr{H}^1(\theta_n \cap \Sigma)\big) \\ &\qquad - A\big(\mathscr{H}^1(\theta_n) - \mathscr{H}^1(\theta_n \cap \Sigma)\big) . \end{aligned} \tag{2.28}$$

Recalling now that $A$ is nondecreasing and continuous, combining (2.27) with (2.26) and (2.28) with (2.25) gives $J(a,b) \le \liminf \delta_\Sigma(\theta_n)$, and therefore $J(a,b) \le \bar{\delta}_\Sigma(\theta)$.

To prove the opposite inequality take $0 \le l \le \mathscr{H}^1(\theta \cap \Sigma)$ and let $\{\theta_n\}$ be, according to Corollary 2.2, a sequence of paths connecting $\theta(0)$ and $\theta(1)$ such that

$$\theta_n \to \theta\,, \quad \mathscr{H}^1(\theta_n) \to \mathscr{H}^1(\theta)\,, \quad \mathscr{H}^1(\theta_n \cap \Sigma) = \mathscr{H}^1(\theta \cap \Sigma) - l \;\forall n \in \mathbb{N}\,.$$

Hence, making use of the continuity of $A$, one gets

$$\delta_\Sigma(\theta_n) = A\big(\mathscr{H}^1(\theta_n) - (\mathscr{H}^1(\theta \cap \Sigma) - l)\big) + B\big(b - l\big) \xrightarrow[n\to\infty]{} A(a+l) + B(b-l)\,.$$

Thus for every $0 \le l \le b$ one has

$$\bar{\delta}_\Sigma(\theta) \le A(a+l) + B(b-l)\,,$$

so the inequality $J(a,b) \ge \bar{\delta}_\Sigma(\theta)$ follows taking the infimum on $l$. □

It is also convenient to introduce an auxiliary function, namely

$$D(a,b) := J(a, b-a)\,; \tag{2.29}$$

indeed, the above proposition tells us that

$$\bar{\delta}_\Sigma(\theta) = J\big(\mathscr{H}^1(\theta \setminus \Sigma), \mathscr{H}^1(\theta \cap \Sigma)\big)\,,$$

or equivalently that

$$\bar{\delta}_\Sigma(\theta) = D\big(\mathscr{H}^1(\theta \setminus \Sigma), \mathscr{H}^1(\theta)\big)\,. \tag{2.30}$$

In other words, we can express $\bar{\delta}_\Sigma(\theta)$ in terms of the length $\mathscr{H}^1(\theta \setminus \Sigma)$ outside of the network and of the length $\mathscr{H}^1(\theta \cap \Sigma)$ inside the network if we make use of $J$, or in terms of the length $\mathscr{H}^1(\theta \setminus \Sigma)$ out of the network and of the total length $\mathscr{H}^1(\theta)$ if we make use of $D$. The advantage of the second possibility, i.e. the advantage of (2.30) with respect to (2.23), is that the variables $\mathscr{H}^1(\theta \setminus \Sigma)$ and $\mathscr{H}^1(\theta)$ satisfy the useful lim inf inequalities (2.25)–(2.26), while the same is not true for $\mathscr{H}^1(\theta \cap \Sigma)$; on the contrary, for $\mathscr{H}^1(\theta \cap \Sigma)$ the lim sup inequality is true, as one can immediately deduce by Lemma 4.1. Another easy interesting property of both $D$ and $J$ is the following one.

**Proposition 2.20.** *The functions $J$ and $D$ are nondecreasing in each of their variables.*

*Proof.* Consider first $J$: take $b \geq 0$ and $a' \geq a \geq 0$; for $0 \leq l \leq b$ one has $A(a+l) \leq A(a'+l)$, so by (2.24) one gets $J(a',b) \geq J(a,b)$ and thus $J$ is nondecreasing in its first variable. Concerning the second one, take $a \geq 0$ and $b' \geq b \geq 0$: one has

$$A(a+l) + B(b'-l) \geq A(a+l) + B(b-l) \geq J(a,b) \qquad \forall\, 0 \leq l \leq b\,;$$

on the other hand, one has

$$A(a+l) + B(b'-l) \geq A(a+b) + B(0) \geq J(a,b) \qquad \forall\, b \leq l \leq b'\,.$$

It follows that $J(a,b') \geq J(a,b)$, so $J$ is nondecreasing also in its second variable.

Consider now $D$: first of all, we rewrite (2.29) in a more convenient way as

$$\begin{aligned} D(a,b) &= \inf\left\{A(a+l) + B(b-a-l) : \ 0 \leq l \leq b-a\right\} \\ &= \inf\left\{A(l) + B(b-l) : \ a \leq l \leq b\right\}. \end{aligned} \tag{2.31}$$

Then, take $b \geq 0$ and $a' \geq a \geq 0$: if $a' \leq l \leq b$ then a fortiori $a \leq l \leq b$, hence one gets $D(a,b) \leq D(a',b)$ directly by (2.31), and consequently $D$ is nondecreasing in its first variable. Finally, concerning the second one, take $a \geq 0$ and $b' \geq b \geq 0$: if $a \leq l \leq b$ then

$$A(l) + B(b'-l) \geq A(l) + B(b-l) \geq D(a,b)\,;$$

on the other hand, if $b \leq l \leq b'$ then

$$A(l) + B(b'-l) \geq A(b) + B(0) \geq D(a,b)\,.$$

It follows that $D(a,b') \geq D(a,b)$, so $D$ is nondecreasing also in its second variable and the proof is completed. □

# Chapter 3
# Optimal Connected Networks

In this chapter we consider the problem of finding an optimal network under the additional constraint that the admissible networks are assumed connected. This extra connectedness assumption enables us to obtain the necessary compactness to ensure the existence of an optimal network. We recall that this case has been extensively studied in [13, 14, 17, 21], where several necessary conditions of optimality have been found; therefore we limit ourselves to recall briefly the main known results. In the last section of this chapter we show some numerical plots of the optimal networks in a slightly simpler situation, where instead of transporting $f^+$ onto $f^-$, the goal of the planner is to transport $f^+$ on the network in the most efficient way.

## 3.1 Optimization Problem

The optimization problem we consider in this chapter is

$$\min\{\mathfrak{F}_L(\Sigma)\ :\ \Sigma \subseteq \Omega,\ \Sigma \text{ connected}\}, \tag{3.1}$$

where $\mathfrak{F}_L$ is the cost functional defined in (2.11) with $H(s) = 0$ for $s \leq L$, $H(s) = \infty$ otherwise. More generally, one may consider the optimization problem

$$\min\{MK(\Sigma)\ :\ \Sigma \subseteq \Omega,\ \Sigma \text{ connected}\} \tag{3.2}$$

where the function $d_\Sigma$ appearing in (2.5) is now given by

$$d_\Sigma(x,y) := \inf\Big\{\Phi\Big(\mathcal{H}^1(\theta\setminus\Sigma), \mathcal{H}^1(\theta\cap\Sigma), \mathcal{H}^1(\Sigma)\Big)\ :\ \theta\in\Theta,\ \theta(0)=x,\ \theta(1)=y\Big\}, \tag{3.3}$$

G. Buttazzo et al., *Optimal Urban Networks via Mass Transportation*,
Lecture Notes in Mathematics 1961, DOI: 10.1007/978-3-540-85799-0_3,

and $\Phi : (\mathbb{R}^+)^3 \to [0, +\infty]$ is a function satisfying the following conditions:

- $\Phi$ is lower semicontinuous;
- $\Phi$ is continuous in its first variable;
- $\Phi$ is nondecreasing in each of its variables, i.e.

$$a_1 \leq a_2,\ b_1 \leq b_2,\ c_1 \leq c_2 \quad \Longrightarrow \quad \Phi(a_1, b_1, c_1) \leq \Phi(a_2, b_2, c_2);$$

- $\Phi(a, b, c) \geq H(c)$ with $H(c) \to \infty$ as $c \to \infty$.

For instance, taking

$$\Phi(a, b, c) := A(a) + B(b) + \chi_L(c)\,,$$

where $\chi_L(c)$ is the function which takes the value zero if $c \leq L$ and $\infty$ otherwise, gives the problem (3.1).

Under the connectedness requirement, the existence of an optimal network always occurs, as the below theorem states.

**Theorem 3.1.** *Under the assumptions above on the function $\Phi$, the optimization problem (3.2) admits a solution $\Sigma_{\text{opt}}$.*

The proof of the above theorem is given in [14]. Here we will sketch its main steps. First of all we equip the admissible class

$$\{\Sigma \subseteq \Omega,\ \Sigma \text{ connected}\}$$

with the topology induced by the Hausdorff distance

$$d_H(\Sigma_1, \Sigma_2) = \sup\{\text{dist}(x_1, \Sigma_2) + \text{dist}(x_2, \Sigma_1)\ :\ x_1 \in \Sigma_1,\ x_2 \in \Sigma_2\}$$

where by $\text{dist}(x, \Sigma)$ we denoted the minimal distance from the point $x$ to the closed set $\Sigma$. It is well-known that the this topology (called Hausdorff topology) is compact and that the Hausdorff limit of a sequence of connected sets is still connected. Moreover, the following result, known as Gołąb theorem (see for instance [5, 37]) holds.

**Theorem 3.2 (Gołąb).** *If $\Sigma_n$ is a sequence of connected closed subsets of $\Omega$ which Hausdorff converges to a set $\Sigma$, then*

$$\mathscr{H}^1(\Sigma) \leq \liminf_{n \to \infty} \mathscr{H}^1(\Sigma_n)\,.$$

*Remark 3.3.* More generally, the above result on lower semicontinuity of the one-dimensional Hausdorff measure $\mathscr{H}^1$ still holds, if we assume that the number of connected components of $\Sigma_n$ is a priori bounded by a fixed constant. On the other hand, it is very easy to find counterexamples which show that the Gołąb theorem fails when the connectedness assumption is completely removed.

By an approximation argument similar to the one we used in the proof of Proposition 2.18 we can compute the relaxed form of the distance $d_\Sigma$ appearing in (3.3). In other words, if convenient, we may approximate a curve $\theta$ passing through the network $\Sigma$ by a sequence of curves $\theta_n$ converging to $\theta$ uniformly and not passing through $\Sigma$. More precisely, the following result holds.

**Proposition 3.4.** *For every closed connected subset $\Sigma$ of $\Omega$ we have*

$$d_\Sigma(x,y) = \inf\Big\{J\Big(\mathcal{H}^1(\theta\setminus\Sigma), \mathcal{H}^1(\theta\cap\Sigma), \mathcal{H}^1(\Sigma)\Big) \; : \; \theta\in\Theta,\ \theta(0)=x,\ \theta(1)=y\Big\}, \tag{3.4}$$

*where the function $J$ is given by*

$$J(a,b,c) := \inf\{\Phi(a+t, b-t, c) \; : \; 0\le t\le b\}.$$

It is now convenient to change variables using $\mathcal{H}^1(\theta)$ instead of $\mathcal{H}^1(\theta\cap\Sigma)$: since

$$\mathcal{H}^1(\theta\cap\Sigma) = \mathcal{H}^1(\theta) - \mathcal{H}^1(\theta\setminus\Sigma),$$

the function $J$ can be replaced by the new one

$$D(a,b,c) = J(a, b-a, c) \tag{3.5}$$

and formula (3.4) becomes

$$d_\Sigma(x,y) = \inf\Big\{D\Big(\mathcal{H}^1(\theta\setminus\Sigma), \mathcal{H}^1(\theta), \mathcal{H}^1(\Sigma)\Big) \; : \; \theta\in\Theta,\ \theta(0)=x,\ \theta(1)=y\Big\}. \tag{3.6}$$

The advantage of using the new function $D$ in (3.5) consists in the following monotonicity and lower semicontinuity property (the proof of the first part is very similar to the one of Proposition 2.20).

**Proposition 3.5.** *The function $D$ defined in (3.5) is monotone nondecreasing and lower semicontinuous in each variable.*

The proof of the existence Theorem 3.1 uses also the following generalization of the Gołąb Theorem, the proof of which can be found in [14], also obtained by Dal Maso and Toader in [30] for different purposes, related to the study of models in fracture mechanics.

**Theorem 3.6.** *Let $X$ be a metric space, $\{\Gamma_n\}_{n\in\mathbb{N}}$ and $\{\Sigma_n\}_{n\in\mathbb{N}}$ be two sequences of compact subsets such that $\Gamma_n\to\Gamma$ and $\Sigma_n\to\Sigma$ in the Hausdorff sense, for some compact subsets $\Gamma$ and $\Sigma$. Let us also suppose that $\Gamma_n$ is connected for all $n\in\mathbb{N}$. Then*

$$\mathscr{H}^1(\Gamma \setminus \Sigma) \leq \liminf_{n\to\infty} \mathscr{H}^1(\Gamma_n \setminus \Sigma_n). \tag{3.7}$$

By using the above extension of the Gołąb theorem and the monotonicity result of Proposition 3.5 it is not difficult to obtain the following lower semicontinuity property for the function $d_\Sigma$.

**Proposition 3.7.** *Let $\{x_n\}_{n\in\mathbb{N}}$ and $\{y_n\}_{n\in\mathbb{N}}$ be sequences in $\Omega$ such that $x_n \to x$ and $y_n \to y$. If $\{\Sigma_n\}_{n\in\mathbb{N}}$ is a sequence of closed connected sets such that $\Sigma_n \to \Sigma$ in the Hausdorff sense, then*

$$d_\Sigma(x,y) \leq \liminf_{n\to\infty} d_{\Sigma_n}(x_n, y_n). \tag{3.8}$$

*In particular, for every closed connected $\Sigma \subseteq \Omega$ the function $d_\Sigma$ is lower semicontinuous on $\Omega \times \Omega$.*

The proof of Theorem 3.1 can be now obtained by putting together the above results. In fact, if $\{\Sigma_n\}$ is a minimizing sequence of the optimization problem (3.2), by the compactness of the Hausdorff convergence we may assume, up to a subsequence, that $\Sigma_n \to \Sigma$ for a suitable closed connected $\Sigma \subseteq \Omega$. The optimality of $\Sigma$ then follows by the lower semicontinuity of the functional $\Sigma \mapsto MK(\Sigma)$ in (3.2) with respect to the Hausdorff convergence. This lower semicontinuity in turn follows by Proposition 3.7 together with the observation that, if $\gamma_n$ are optimal plans with respect to $\Sigma_n$ and $\gamma_n \overset{*}{\rightharpoonup} \gamma$, then $\gamma$ is an optimal plan with respect to $\Sigma$. All the details of the argument above can be found in the paper [14], to which we refer the interested reader.

## 3.2 Properties of the Optimal Networks

In this section we collect the results of [19, 20, 54] regarding the case of optimization problem (3.1) in the particular case when $A(t) = t$, $B(t) = 0$ and $\Sigma$ is *a priori* required to be *connected.* Since the results we present here are not used elsewhere in this monograph, we give only the outlines of the proofs of some of the most important results. For the details the reader is referred to the aforementioned papers.

First, we recall the following basic results regarding topological structure and regularity of optimal sets.

**Theorem 3.8.** *Let in problem* (3.1) $A(t) = t$ *and* $B(t) = 0$, *and assume that* $f^+, f^- \ll \mathscr{L}^N$ *and* $f^+ \neq f^-$. *Then every closed connected* $\Sigma \subseteq \mathbb{R}^N$ *solving this problem has the following properties.*

- $\mathscr{H}^1(\Sigma) = L$.
- *$\Sigma$ contains no closed curve (a homeomorphic image of $S^1$). In particular, $\mathbb{R}^N \setminus \Sigma$ is connected.*

*If, moreover, $f^+$, $f^- \in L^2(\Omega)$, then the following additional properties hold.*

- *$\Sigma$ is Ahlfors regular in the sense that for every $x \in \Sigma$ and $r < \operatorname{diam}\Sigma$ one has*

$$cr \leq \mathcal{H}^1(\Sigma \cap B_r(x)) \leq Cr$$

*for some constants $c > 0$ and $C > 0$ independent both on $x$ and on $r > 0$.*

*Remark 3.9.* The results of the papers [19, 20, 54] are in fact a bit sharper than what claimed above. For instance the first two statements are valid even if the dimension of the measures $f^+$, $f^-$ (defined in a suitable way) is sufficiently high. Further, the Ahlfors regularity of optimal sets is known to hold when $f^+$, $f^- \in L^p(\Omega)$ where $p = 4/3$ when $N = 2$ and $p = N/(N-1)$ otherwise (although it is not known whether such assumption is sharp).

We provide here just the general outline of the proof of the first two statements above omitting technical details. The principal technical tool is given by the proposition below in which a set $\Sigma$ is modified with the addition of a piece of small length to obtain a new set such that all points which were not too close to $\Sigma$ become closer to the modified set (see Proposition 4.1 from [54]).

**Proposition 3.10.** *Let $\Sigma \subseteq \mathbb{R}^N$ be a compact connected set with $\mathcal{H}^1(\Sigma) < +\infty$ and $r > 0$ be some given number. Then for each $c > 0$ there exists a compact connected $\Sigma' \subseteq \mathbb{R}^N$, $\Sigma' \supseteq \Sigma$, with $\mathcal{H}^1(\Sigma') \leq \mathcal{H}^1(\Sigma) + c$ and such that for every $y \in \mathbb{R}^N$ satisfying $\operatorname{dist}(y, \Sigma) \geq r$ one has*

$$\operatorname{dist}(y, \Sigma') \leq \operatorname{dist}(y, \Sigma) - C,$$

*where $C > 0$ is some constant depending on $\Sigma$, $r$, $c$ and the space dimension $N$ but independent of $y$.*

Now, with the help of the above Proposition 3.10, the proof of the first assertion of Theorem 3.8 becomes quite simple.

*Outline of the proof of $\mathcal{H}^1(\Sigma) = L$.* Suppose the contrary, i.e. that $\mathcal{H}^1(\Sigma) = L - c$ for some $c > 0$, where $\Sigma$ is a connected set solving problem (3.1). Let now $\Gamma \subseteq \mathbb{R}^N \times \mathbb{R}^N$ be the set of those pairs $(x, y)$ such that $d_\Sigma(x, y) < d_\emptyset(x, y)$, hence the segment $[x, y]$ is not a geodesic between $x$ and $y$. If now $\gamma$ is an optimal transport plan, we set $\varphi$ the measure

$$\varphi := \pi_{1\#}(\gamma \llcorner \Gamma) + \pi_{2\#}(\gamma \llcorner \Gamma)\,.$$

It is easy to show that then $\Sigma$ minimizes the average distance functional

$$\Sigma \mapsto F_\varphi(\Sigma) := \int_\Omega \operatorname{dist}(x, \Sigma)\, d\varphi(x)$$

over closed connected sets satisfying the length constraint $\mathcal{H}^1(\Sigma) \leq L$. We choose an $r > 0$ such that

$$\varphi(D_r) > 0, \qquad \text{where } D_r := \{y \in \mathbb{R}^N : \text{dist}\,(y, \Sigma) > r\}.$$

Such an $r > 0$ exists since otherwise $\varphi$ would be concentrated over $\Sigma$, which cannot happen since $\varphi \leq f^+ + f^-$ and $f^+ + f^- \ll \mathscr{L}^N$. Consider now a set $\Sigma'$ provided by Proposition 3.10, so that $\mathscr{H}^1(\Sigma') \leq L$. Besides, for every $y \in \mathbb{R}^N$ one has

$$\text{dist}\,(y, \Sigma') \leq \text{dist}\,(y, \Sigma)$$

since $\Sigma \subseteq \Sigma'$, while

$$\text{dist}\,(y, \Sigma') \leq \text{dist}\,(y, \Sigma) - C$$

for some $C > 0$ (independent of $y$) whenever $y \in D_r$. Hence, minding the strict monotonicity of $A$, we get

$$\begin{aligned} F_\varphi(\Sigma') &= \int_{\mathbb{R}^N \setminus D_r} \text{dist}\,(y, \Sigma')\, d\varphi(y) + \int_{D_r} \text{dist}\,(y, \Sigma')\, d\varphi(y) \\ &\leq \int_{\mathbb{R}^N \setminus D_r} \text{dist}\,(y, \Sigma)\, d\varphi(y) + \int_{D_r} (\text{dist}\,(y, \Sigma) - C)\, d\varphi(y) \\ &< \int_{\mathbb{R}^N \setminus D_r} \text{dist}\,(y, \Sigma)\, d\varphi(y) + \int_{D_r} \text{dist}\,(y, \Sigma)\, d\varphi(y) \\ &= F_\varphi(\Sigma), \end{aligned}$$

contradicting the optimality of $\Sigma$. □

To outline the proof of the fact that optimal connected sets do not contain closed curves (homeomorphic images of $S^1$), we recall briefly the following topological notions which will be used in the sequel.

**Definition 3.11.** Let $\Sigma$ be a connected set. Then $x \in \Sigma$ is called *noncut point* of $\Sigma$, if $\Sigma \setminus \{x\}$ is connected. Otherwise, $x$ is called *cut point* of $\Sigma$.

Let us recall that according to the Moore theorem (Theorem IV.5 from [46, § 47]), every continuum (i.e. compact connected space) has at least two noncut points. We also need the following statement from [21].

**Lemma 3.12.** *Let $\Sigma \subseteq \mathbb{R}^N$ be a closed connected set satisfying $\mathscr{H}^1(\Sigma) < +\infty$ which contains a closed curve $S$. Then $\mathscr{H}^1$-a.e. point $x \in S$ is a noncut point for $\Sigma$.*

With the help of the above Lemma 3.12 we are able to prove the following quantitative result (see Lemma 5.4 from [54]), which says that if $\Sigma$ contains a closed curve, one can almost everywhere take away a small piece of it such that the remaining part is still connected (it is here that Lemma 3.12 is invoked), while the increase in average distance functional $F_\varphi$ is small (with the quantitative estimate).

**Lemma 3.13.** *Suppose that $\Sigma \subseteq \mathbb{R}^N$ is a compact connected set satisfying $\mathcal{H}^1(\Sigma) < +\infty$ and containing a closed curve $S$. Given $\beta \in (0,1]$ and $r > 0$, for $\mathcal{H}^1$-a.e. $\bar{x} \in S$ there exists a $\rho \in (0,r)$ and a closed connected set $\Sigma' \subseteq \mathbb{R}^N$ such that*

- $\mathcal{H}^1(\Sigma') \leq \mathcal{H}^1(\Sigma) - \rho/2 + C_2\beta\rho$,
- $\Sigma \setminus \Sigma' \subseteq B(\bar{x}, \rho)$,
- $\Sigma' \setminus \Sigma \subseteq B(\bar{x}, 32N\rho)$,
- $\operatorname{dist}(y, \Sigma') \leq \operatorname{dist}(y, \Sigma)$, *for all* $y \notin B(\bar{x}, 64N\sqrt{N}\rho)$,
- $\operatorname{dist}(y, \Sigma') \leq \operatorname{dist}(y, \Sigma) + \rho$, *for all* $y \in B(\bar{x}, 64N\sqrt{N}\rho)$,

*where $C_2 > 0$ is a constant depending only on $N$.*

*Outline of the proof of absence of loops in optimal sets.*

Let again $\Sigma$ be a solution to (3.1) (hence a closed connected set minimizing $F_\varphi$). Since $\varphi(\Sigma) = 0$, there exists a compact set $K$, disjoint from $\Sigma$ and such that $\varphi(K) > 0$. Let

$$R := \frac{1}{2}\min\Big\{\operatorname{dist}(y, \Sigma) : y \in K\Big\} > 0$$

and let $H$ be the $R$-neighborhood of $\Sigma$.

Suppose by contradiction that there exists a simple closed curve $S \subseteq \Sigma$. Given $x \in \Sigma$, define

$$\omega(x, \rho) := \frac{\varphi(B(x, \rho))}{\rho}.$$

Since $\varphi(S) = 0$, as direct consequence of [2, Theorem 2.56], for $\mathcal{H}^1$-a.e. $x \in S$ one has

$$\lim_{\rho \to 0^+} \omega(x, \rho) = 0.$$

Let $r > 0$ to be chosen later. We apply therefore Lemma 3.13 with $\beta$ small enough to find a point $\bar{x} \in S$ with $\omega(x,t) \to 0$ as $t \to 0^+$, a $\rho \in (0,r)$ and a connected set $\Sigma'$ such that

$$\mathcal{H}^1(\Sigma') \leq \mathcal{H}^1(\Sigma) - \rho/2 + C_2\beta\rho \leq \mathcal{H}^1(\Sigma) - \rho/4, \tag{3.9}$$

while

$$\begin{aligned} F_\varphi(\Sigma') &\leq F_\varphi(\Sigma) + \rho\varphi(B(\bar{x}, 64N\sqrt{N}\rho)) \\ &\leq F_\varphi(\Sigma) + 64N\sqrt{N}\rho^2\omega(\bar{x}, 64N\sqrt{N}\rho). \end{aligned} \tag{3.10}$$

We now need the following geometrical result, see [54, Lemma 3.5].

**Lemma 3.14.** *Let $l > 0$ be given and let $H$ and $K$ be two Borel subsets of $\mathbb{R}^N$ such that $\varphi(K) > 0$ and*

$$s := \inf\{\operatorname{dist}(y, H) : y \in K\} > 0.$$

*Then given any compact connected set $\Delta \subseteq H$ with $\mathscr{H}^1(\Delta) \leq l$, one has for all $\varepsilon < s/\sqrt{N}$ the existence of a compact connected set $\Delta' \supseteq \Delta$ such that*

$$\begin{aligned} \mathscr{H}^1(\Delta') &\leq \mathscr{H}^1(\Delta) + 2N\varepsilon, \\ F_\varphi(\Delta') &\leq F_\varphi(\Delta) - C_1\varepsilon^2, \end{aligned} \tag{3.11}$$

*where*

$$C_1 := \frac{\varphi(K)}{32Nl}.$$

Applying Lemma 3.14 with $\varepsilon := \rho/8N$ (which is admissible provided $r$, and then $\rho$, is small enough), $\Delta = \Sigma'$, $s = R$ and $l = L$ (mind that $\mathscr{H}^1(\Sigma') \leq \mathscr{H}^1(\Sigma) = L$), we find a connected compact set $\Sigma'' := \Delta' \supseteq \Sigma'$ such that, by (3.9)

$$\mathscr{H}^1(\Sigma'') \leq \mathscr{H}^1(\Sigma') + 2N\varepsilon \leq \mathscr{H}^1(\Sigma) - \rho/4 + 2N\varepsilon = \mathscr{H}^1(\Sigma),$$

while, by (3.11) and (3.10),

$$\begin{aligned} F_\varphi(\Sigma'') &\leq F_\varphi(\Sigma') - C_1\varepsilon^2 \\ &\leq F_\varphi(\Sigma) + 64N\sqrt{N}\rho^2\omega(\bar{x}, 64N\sqrt{N}\rho) - \frac{C_1}{64N^2}\rho^2 \\ &= F_\varphi(\Sigma) - C\rho^2 + o(\rho^2), \end{aligned} \tag{3.12}$$

where $C_1$ is the constant introduced in Lemma 3.14, and $C := C_1/64N^2$. Hence, choosing a sufficiently small $r > 0$ and minding that $\rho \leq r$, we get from (3.12) that $F_\varphi(\Sigma'') < F_\varphi(\Sigma)$, contradicting the optimality of $\Sigma$. □

To discuss deeper properties of optimal sets, we need to recall some topological notions from [46].

**Definition 3.15.** Let $\Sigma$ be a topological space. We will say that *the order of the point $x \in \Sigma$ does not exceed* $\mathfrak{n}$, writing $\mathrm{ord}_x\Sigma \leq \mathfrak{n}$, where $\mathfrak{n}$ is a cardinal, if for every $\varepsilon > 0$ there is an open subset $U \subseteq \Sigma$ such that $x \in U$, $\mathrm{diam}\,(U) < \varepsilon$ and $\#\partial U \leq \mathfrak{n}$, # standing for the cardinality of a set.

The *order of the point $x \in \Sigma$* is said to be $\mathfrak{n}$, written $\mathrm{ord}_x\Sigma = \mathfrak{n}$, if $\mathfrak{n}$ is the least cardinal for which $\mathrm{ord}_x\Sigma \leq \mathfrak{n}$.

If $\mathrm{ord}_x\Sigma = \mathfrak{n}$, with $\mathfrak{n} \geq 3$, then $x$ will be called a *branching point* of $\Sigma$, while if $\mathrm{ord}_x\Sigma = 1$, then $x$ will be called an *endpoint* of $\Sigma$.

The following results hold (see [19, 63, 66]).

**Theorem 3.16.** *Let in problem* (3.1) $N = 2$, $A(t) = t$ *and* $B(t) = 0$, *and assume that* $f^+, f^- \in L^{4/3}(\Omega)$, *with* $f^+ \neq f^-$. *Then every closed connected* $\Sigma \subseteq \mathbb{R}^N$ *solving the problem has the following properties.*

- $\mathrm{ord}_x\Sigma \leq 3$ *for all* $x \in \Sigma$, *and, further, the number of endpoints and that of branching points (which hence are all triple points) is finite.*

- *the generalized mean curvature* $\mathcal{H}$ *of* $\Sigma$, *defined to be the vector-valued distribution*

$$\langle \mathcal{H}, X \rangle := \int_\Sigma div^\Sigma X \, d\mathscr{H}^0 \qquad \text{for all } X \in C_0^\infty(\mathbb{R}^N, \mathbb{R}^N),$$

*is a Radon measure and* $\mathcal{H}(\{x\}) = 0$ *whenever* $x \in \Sigma$ *is a branching point. Here* $div^\Sigma$ *stands for the tangential divergence operator with respect to* $\Sigma$ *(see [2]), i.e. in other words, the projection of the usual divergence on the approximate tangent space to* $\Sigma$, *the latter defined* $\mathscr{H}^1$*-a.e. on* $\Sigma$. *This property may be interpreted as a "weak form" of the assertion that every branching point is a "regular tripod", i.e. a triple point where three smooth branches meet with angles of 120 degrees.*
- *for any point* $x \in \Sigma$ *there is a number* $r_0 > 0$ *such that for any arc* $\theta \subseteq \Sigma$ *starting at* $x$ *we have*

$$\#\theta \cap \partial B(x, r) = 1$$

*for any* $r \in (0, r_0]$.
- *If* $f^\pm \in L^\infty(\Omega)$, *then for every branching point* $x \in \Sigma$ *there is* $\delta > 0$ *such that the set* $\Sigma \cap B_\delta(x)$ *consists of exactly three* $C^{1,1}$*-arcs starting at* $x$.
- *If* $f^\pm \in L^\infty(\Omega)$ *and* $x \in \Sigma$ *is not a branching point, then the Hausdorff dimension of the set* $k^{-1}(x)$ *is at least* 1, *where* $k : \mathbb{R}^2 \to \Sigma$ *stands for the projection map on* $\Sigma$ *defined a.e. on* $\mathbb{R}^2$.

Since the proof of the above statement is quite involved and is not the principal subject of this monograph, we only give few of its basic ideas. In fact, to prove that the number of endpoints (hence that of branching points) is finite, one shows that for every endpoint $z \in \Sigma$ the mass $\psi(\{z\})$ transported to to it is nonzero, and, moreover, $\psi(\{z\}) > c$ for some $c > 0$ independent of $z$. This is achieved by contradiction using an argument vaguely similar to that of Section 6.1 in the sequel. Once one knows that the number of branching points is finite, one proves that the latter are, in a certain sense regular tripods, by means of comparing $\Sigma$ with $\Sigma'$, where $\Sigma'$ is obtained by substituting, in a sufficiently small neighborhood of a branching point, the part of $\Sigma$ with the Steiner minimal tree connecting points of $\Sigma$ on the boundary of this neighborhood. The rest of the results are obtained by fine blow-up arguments. For the details the reader is referred to [19, 63, 66].

## 3.3 Average Distance Problem

We consider in this section a slightly simpler situation which occurs when, instead of transporting $f^+$ onto $f^-$, the goal of the planner is to transport $f^+$ on $\Sigma$ in the most efficient way. This corresponds to minimizing the total cost over all the measures $f^-$ concentrated on $\Sigma$ and with $\|f^-\| = \|f^+\|$; thus, denoting by $MK(\Sigma, f^+, f^-)$ the cost of transportation of $f^+$ onto $f^-$ with network $\Sigma$, introduced in (2.9), the problem is simply

$$\min\left\{MK(\Sigma, f^+, f^-) \,:\, f^- \in \mathcal{M}^+(\Omega),\ \|f^-\| = \|f^+\|,\ \operatorname{spt} f^- \subseteq \Sigma\right\}.$$

We denote the latter minimum value as $M(\Sigma)$. We are interested in minimizing the quantity $M(\Sigma)$ over all admissible sets $\Sigma$. In particular, when $A(t) = t$, this leads to the so called *average distance problem*, formulated as follows:

$$\min\left\{\int_\Omega \operatorname{dist}(x, \Sigma) f^+(x)\, dx \,:\, \Sigma \subseteq \Omega,\ \Sigma \text{ connected},\ \mathcal{H}^1(\Sigma) \le L)\right\},$$

which corresponds to finding a network $\Sigma_{\rm opt}$ for which the *average distance* for a citizen to reach the closest point of $\Sigma_{\rm opt}$ is minimal. Notice that, since there is no "movement" on $\Sigma$, the above problem does not depend on the choice of the function $B$. Problems with a similar functional, with a fixed set $\Sigma$ also intervene in the study of equilibrium configurations of growing sandpiles (see for instance [26]).

Although all the results of Section 3.2 cannot be just formally applied to this problem, the techniques of the proof still work, and hence, the existence result and the necessary conditions of optimality from Section 3.2 still hold.

We provide some numerical plots of the optimal network $\Sigma_{\rm opt}$ in the average distance case. In Figures 3.1 and 3.2 the set $\Omega$ is the unit square in $\mathbb{R}^2$ and $f^+ \equiv 1$, while in Figures 3.3 and 3.4 the set $\Omega$ is the unit ball in $\mathbb{R}^3$ and again $f^+ \equiv 1$ (the dark dot is the center of the ball). All these plots are taken from [17] and, since the problem has a great number of local minima, have been obtained by using an Evolutionary Algorithm (EAs) with an adaptive penalty method. For more details we refer the reader to [17].

We conclude this section by pointing out the question of studying the regularity of optimal solutions $\Sigma_{\rm opt}$ under the connectedness assumption.

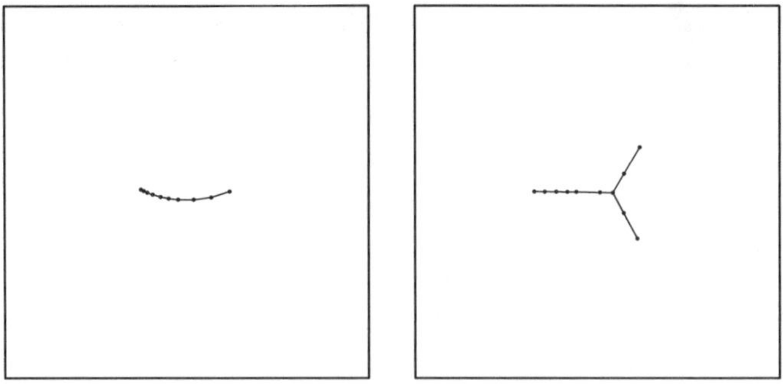

**Fig. 3.1** Plot of $\Sigma_{\rm opt}$ with $L = 0.25$ (left) and $L = 0.5$ (right).

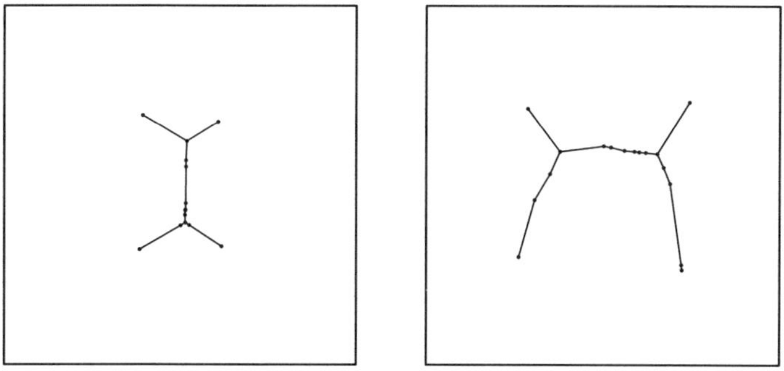

**Fig. 3.2** Plot of $\Sigma_{\text{opt}}$ with $L = 0.75$ (left) and $L = 1.25$ (right).

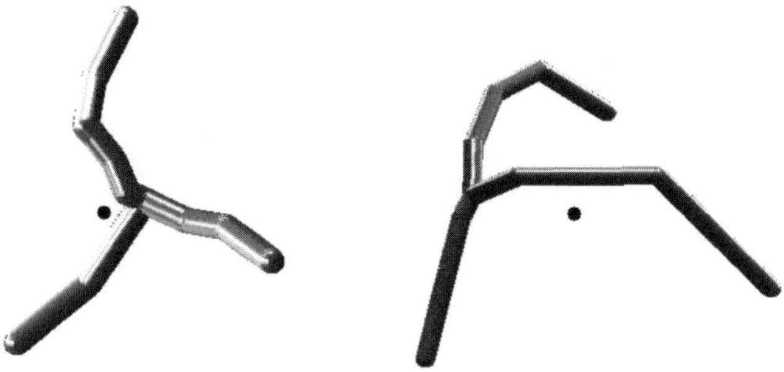

**Fig. 3.3** Plot of $\Sigma_{\text{opt}}$ with $L = 0.25$ (left) and $L = 0.5$ (right).

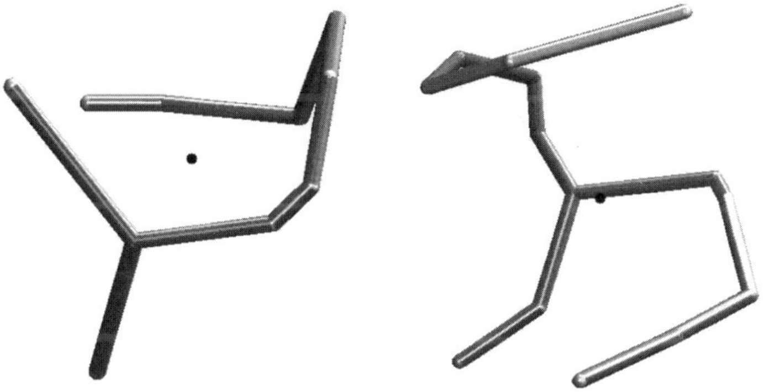

**Fig. 3.4** Plot of $\Sigma_{\text{opt}}$ with $L = 0.25$ (left) and $L = 0.5$ (right).

Even if we assume that all the data are extremely smooth, this revealed to be a rather difficult issue, and very little is known, apart the Ahlfors regularity of minimizers provided by Theorem 3.8 as well as the rather weak regularity result given by Theorem 3.16. Nevertheless, one may expect that every minimizer $\Sigma_{\text{opt}}$ is a finite union of (at least) $C^1$ curves; in other words, it is meaningful to conjecture that $\Sigma_{\mathrm{opt}}$ is (at least) $C^1$ regular outside a finite number of branching points, any of which, at least in the case $N = 2$, is a 120 degrees regular tripod. For a more detailed discussion of the problem, as well as for a list of open problems and conjectures regarding the case of optimal connected networks, we refer the interested reader to [17, 19, 20].

# Chapter 4
# Relaxed Problem and Existence of Solutions

In the general setting that we are considering, the existence of solutions (i.e. Borel sets $\Sigma$ minimizing the functional $\mathfrak{F}$ defined in (2.11)) may fail, as we will see in the example of Section 4.3; this is why we introduce a relaxed version of the problem. This relaxation is very convenient for a number of reasons: first of all it deals, instead of Borel sets, with Radon measures, which have a considerably better structure and nicer properties; moreover, it easily admits solutions and, as we will see, it is possible to study these solutions and to understand whether or not they are classical, i.e. of the form $\mathscr{H}^1 \llcorner \Sigma$ for a set $\Sigma$. Finally, we will give an interpretation of the meaning of the relaxed problem, and generalize all the definitions that we presented in Chapter 2 for the classical setting. In Section 4.1 we introduce the relaxed optimization problem and we show that it always admits a solution. In Section 4.2 we show some properties of the relaxed solutions, in particular that there is a solution $\mu \in \mathcal{M}^+(\Omega)$ such that for some rectifiable set $\Sigma$ one has $\mu = a(x)\mathscr{H}^1 \llcorner \Sigma$, where $a : \Sigma \to [0, 1]$ is some density; the relaxed solution $\mu$ is then classical if and only if $a(x) \equiv 1$ for $\mu$−a.e. $x \in \Omega$ and, in this case, $\mu$ corresponds to the rectifiable set $\{x : a(x) = 1\}$. In Section 4.3 we present a class of examples in which all the solutions are not classical. Finally, in Section 4.4 we show that there are classical solutions if the function $D$ defined in (2.29) is concave in the first variable, and that all the solutions are classical if this concavity is strict.

## 4.1 Relaxed Problem Setting

First of all, we need to recall an easy abstract result.

**Lemma 4.1.** *Let $\{C_n\}$ be a sequence of closed sets of some metric space $X$ converging in the Hausdorff distance to $C$. Let moreover $\{\nu_n\} \subseteq \mathcal{M}^+(X)$ be a sequence of measures weakly* converging to $\nu \in \mathcal{M}^+(X)$. Then, one has*

G. Buttazzo et al., *Optimal Urban Networks via Mass Transportation*, Lecture Notes in Mathematics 1961, DOI: 10.1007/978-3-540-85799-0_4,

$$\nu(C) \geq \limsup \nu_n(C_n) .$$

*Proof.* Since $\{C_n\}$ converges in the Hausdorff distance to $C$, having fixed any $\varepsilon > 0$ one has $C_n \subseteq B_\varepsilon(C)$ for $n$ large enough, where

$$B_\varepsilon(C) = \{x \in X : d_X(x, C) \leq \varepsilon\} ;$$

therefore, since $B_\varepsilon(C)$ is a closed set, one has

$$\nu\big(B_\varepsilon(C)\big) \geq \limsup_{n\to\infty} \nu_n\big(B_\varepsilon(C)\big) \geq \limsup_{n\to\infty} \nu_n(C_n) .$$

The proof is completed noticing that $\nu\big(B_\varepsilon(C)\big) \searrow \nu(C)$ for $\varepsilon \searrow 0$. □

We now notice a couple of properties of $d_\Sigma$ and $\mathfrak{F}$: the first one is the lower semicontinuity of $d_\Sigma$ with respect to the weak* convergence of the measures $\mathscr{H}^1 \llcorner \Sigma$; as a consequence, we can also show the lower semicontinuity of the functional $\mathfrak{F}$ with respect to the same convergence. Before showing these properties, we define formally the convergence we are going to use.

**Definition 4.2.** Given a sequence $\{\Sigma_n\}$ of Borel sets of bounded length, we say that $\{\Sigma_n\}$ *converges in the* $\mathscr{H}^1$ *sense to* $\Sigma$, shortly $\Sigma_n \xrightarrow{\mathscr{H}^1} \Sigma$, if the measures $\mathscr{H}^1 \llcorner \Sigma_n$ weakly* converge to the measure $\mathscr{H}^1 \llcorner \Sigma$ in $\mathcal{M}^+(\Omega)$.

**Proposition 4.3.** *Let* $\{\Sigma_n\} \xrightarrow{\mathscr{H}^1} \Sigma$ *and* $\theta_n \to \theta$ *uniformly: then*

$$\bar{\delta}_\Sigma(\theta) \leq \liminf_{n\to\infty} \bar{\delta}_{\Sigma_n}(\theta_n) .$$

*Moreover, for each* $x, y \in \Omega$

$$d_\Sigma(x, y) \leq \liminf_{n\to\infty} d_{\Sigma_n}(x, y) .$$

*Proof.* Applying Lemma 4.1 with $C_n = \theta_n$, $C = \theta$, $\nu_n = \mathscr{H}^1 \llcorner \Sigma_n$ and $\nu = \mathscr{H}^1 \llcorner \Sigma$, one obtains

$$\mathscr{H}^1(\theta \cap \Sigma) \geq \limsup \mathscr{H}^1(\theta_n \cap \Sigma_n) .$$

Therefore, setting

$$a := \mathscr{H}^1(\theta \setminus \Sigma) , \quad b := \mathscr{H}^1(\theta \cap \Sigma) , \quad l_n := \mathscr{H}^1(\theta \cap \Sigma) - \mathscr{H}^1(\theta_n \cap \Sigma_n) ,$$

one has that $l_n \leq b$ and $\liminf l_n \geq 0$. Recalling then formulas (2.23) and (2.24) for $\bar{\delta}_\Sigma$, and the facts that $A$ and $B$ are nondecreasing, we obtain

$$
\begin{aligned}
\liminf_{n\to\infty} \bar\delta_{\Sigma_n}(\theta_n) &= \liminf_{n\to\infty} A\big(\mathscr{H}^1(\theta_n) - \mathscr{H}^1(\theta_n\cap\Sigma_n)\big) + B\big(\mathscr{H}^1(\theta_n\cap\Sigma_n)\big)\\
&= \liminf_{n\to\infty} A\Big(\mathscr{H}^1(\theta_n) - \mathscr{H}^1(\theta_n\cap\Sigma_n) + \mathscr{H}^1(\theta\cap\Sigma) + \mathscr{H}^1(\theta\setminus\Sigma) - \mathscr{H}^1(\theta)\Big)\\
&\qquad + B\Big(\mathscr{H}^1(\theta\cap\Sigma) - \big(\mathscr{H}^1(\theta\cap\Sigma) - \mathscr{H}^1(\theta_n\cap\Sigma_n)\big)\Big)\\
&= \liminf_{n\to\infty} A\big(a + l_n + \mathscr{H}^1(\theta_n) - \mathscr{H}^1(\theta)\big) + B\big(b - l_n\big)\\
&\geq \liminf_{n\to\infty} A\big(a + l_n\big) + B\big(b - l_n\big)\\
&\geq \inf\big\{A\big(a+l\big) + B\big(b-l\big) \,:\, 0\leq l\leq b\big\} = J(a,b) = \bar\delta_\Sigma(\theta)\,,
\end{aligned}
$$

where the first inequality is due to the fact that $a+l_n$ is bounded, and hence $A$ is uniformly continuous, and to the inequality

$$\mathscr{H}^1(\theta) \leq \liminf \mathscr{H}^1(\theta_n)\,.$$

Thus the first part of the claim follows. Concerning the second part, we argue by contradiction: if it were not true that $d_\Sigma \leq \liminf d_{\Sigma_n}$, then there would exist a pair $x$, $y\in\Omega$ and $\varepsilon > 0$ such that

$$d_{\Sigma_n}(x,y) \leq d_\Sigma(x,y) - \varepsilon$$

for countably many $n\in\mathbb{N}$. According to Corollary 2.11 and Lemma 2.16, for any $n$ we select a path $\theta_n$ with

$$\mathscr{H}^1(\theta_n) \leq L\,, \qquad\qquad \bar\delta_{\Sigma_n}(\theta_n) = d_{\Sigma_n}(x,y)\,.$$

Hence, up to a subsequence we can assume that $\theta_n \to \theta$ uniformly and that

$$\bar\delta_{\Sigma_n}(\theta_n) \leq d_\Sigma(x,y) - \varepsilon$$

for any $n$. The first part of the proof ensures that

$$\bar\delta_\Sigma(\theta) \leq d_\Sigma(x,y) - \varepsilon < d_\Sigma(x,y)\,,$$

which gives the desired contradiction and so also the second part of the claim follows. □

The assertion below is a consequence of the previous proposition.

**Corollary 4.4.** *The functional $\mathfrak{F}$ is l.s.c. with respect the convergence in the $\mathscr{H}^1$ sense.*

*Proof.* If $\{\Sigma_n\} \xrightarrow{\mathscr{H}} \Sigma$, then

$$\mathscr{H}^1(\Sigma) = \lim \mathscr{H}^1(\Sigma_n) < +\infty\,.$$

Further, by Proposition 4.3 one has

$$\bar{\delta}_\Sigma \leq \Gamma - \liminf \bar{\delta}_{\Sigma_n},$$

and hence, by Lemma B.20 of the Appendix one has

$$\overline{C}_\Sigma(\eta) \leq \liminf \overline{C}_{\Sigma_n}(\eta_n)$$

whenever $\eta_n \xrightarrow{*} \eta$ in the weak* sense of measures. According to Proposition 2.14, we may take $\eta_n$ such that $MK(\Sigma_n) = \overline{C}_{\Sigma_n}(\eta_n)$; since by Corollary 2.17 one may assume that the supports of $\eta_n$ are bounded in $\Theta$ by some constant $L$ independent of $n$, minding that

$$\|\eta_n\|_{\mathcal{M}^+(\Theta)} = \|f^+\|_{\mathcal{M}^+(\Omega)} < +\infty,$$

we get $\eta_n \xrightarrow{*} \eta$ for some t.p.m. $\eta$ up to a subsequence. Hence

$$MK(\Sigma) \leq \overline{C}_\Sigma(\eta) \leq \liminf_{n\to\infty} \overline{C}_{\Sigma_n}(\eta_n) = \liminf_{n\to\infty} MK(\Sigma_n);$$

moreover, from (2.10) we deduce

$$H(\mathscr{H}^1(\Sigma)) \leq \liminf H\big(\mathscr{H}^1(\Sigma_n)\big)$$

and finally, by the definition (2.11) of the functional $\mathfrak{F}$, we deduce

$$\mathfrak{F}(\Sigma) \leq \liminf \mathfrak{F}(\Sigma_n).$$

□

Let us finally introduce the relaxed version of the problem: we consider as relaxed admissible networks all the nonnegative measures $\mu \in \mathcal{M}^+(\Omega)$, identifying any classical network $\Sigma$ with the measure $\mathscr{H}^1 \llcorner \Sigma$. In a standard way, we define then the relaxed functional (using the same symbol with a slight abuse of notation) as

$$\mathfrak{F}(\mu) := \inf\left\{\liminf_{n\to\infty} \mathfrak{F}(\Sigma_n) : \mathscr{H}^1 \llcorner \Sigma_n \xrightarrow{*} \mu\right\}, \tag{4.1}$$

so that the relaxed optimization problem now reads

$$\min\{\mathfrak{F}(\mu) : \mu \in \mathcal{M}^+(\Omega)\}. \tag{4.2}$$

It is worth remarking that, thanks to the lower semicontinuity proved in Corollary 4.4, one has $\mathfrak{F}(\Sigma) = \mathfrak{F}(\mathscr{H}^1 \llcorner \Sigma)$, which justifies the above abuse of notations.

We list now some properties of the relaxed functional $\mathfrak{F}$.

**Proposition 4.5.** *The following three properties hold:*

i) *$\mathfrak{F}$ is coercive, that is, $\mathfrak{F}(\mu_n) \to \infty$ if $\|\mu_n\| \to \infty$;*

ii) $\mathfrak{F}$ *is l.s.c. in* $\mathcal{M}^+(\Omega)$ *with respect to the weak* convergence;*

iii) $\min\{\mathfrak{F}(\mu) : \mu \in \mathcal{M}^+(\Omega)\} =$
$\inf\{\mathfrak{F}(\Sigma) : \Sigma$ *is a Borel set of finite length*$\}$.

*Proof.* This is a standard fact in relaxation theory, which follows for instance from Proposition 1.3.1 and 1.3.5 of [16]. □

## 4.2 Properties of Relaxed Minimizers

In this section we show some properties of an optimal measure $\mu$, which exists thanks to Proposition 4.5; first of all, we establish the existence of an upper bound for the length of the paths contained in the support of optimal t.p.m.'s related to a minimizing sequence of sets $\Sigma_n$.

**Lemma 4.6.** *There is a constant* $L \geq 0$, *depending only on* $A$, $\Omega$ *and* $H$, *such that the following holds: given any minimizing sequence* $\{\Sigma_n\}$ *for* $\mathfrak{F}$, *for any* $n \in \mathbb{N}$ *sufficiently large there is a t.p.m.* $\eta_n$, *optimal for the functional* $\overline{C}_{\Sigma_n}$, *whose support is bounded in* $\Theta$ *by* $L$. *It is also possible to find a sequence of t.p.m.* $\{\eta_n\}$ *with supports bounded in* $\Theta$ *by* $L$, *such that*

$$C_{\Sigma_n}(\eta_n) \leq MK(\Sigma_n) + \varepsilon_n$$

*with* $\varepsilon_n \to 0$ *(in the following, such a sequence is referred to as "almost optimal with respect to* $C_{\Sigma_n}$*").*

*Proof.* Keeping in mind Corollary 2.17, to show the first assertion it is enough to check that, for each sequence $\{\Sigma_n\}$ minimizing $\mathfrak{F}$ and $n$ large enough, the length $\mathcal{H}^1(\Sigma_n)$ is bounded by a constant depending only on $A$, $\Omega$ and $H$. To this aim, take $l \geq 0$ such that

$$H(l) - 1 \geq MK(\emptyset) = \mathfrak{F}(\emptyset),$$

and notice that $l$ depends only on $H, \Omega$ and $A$ but not on $B$. If $\mathcal{H}^1(\Sigma) \geq l$, one has

$$\mathfrak{F}(\Sigma) \geq H(\mathcal{H}^1(\Sigma)) \geq H(l) \geq \mathfrak{F}(\emptyset) + 1 \geq \min\{\mathfrak{F}(\mu) : \mu \in \mathcal{M}^+(\Omega)\} + 1.$$

Therefore, if $\{\Sigma_n\}$ is a sequence minimizing $\mathfrak{F}$, then $\mathcal{H}^1(\Sigma_n) < l$ for $n$ large enough.

To show the second assertion for a sequence of t.p.m.'s almost optimal with respect to $C_{\Sigma_n}$, we apply Lemma 2.13: taking, for each $n \in \mathbb{N}$, a t.p.m. $\eta_n$ with support bounded in $\Theta$ by $L$ and optimal with respect to $\overline{C}_{\Sigma_n}$, the t.p.m.'a $\alpha_{\varepsilon_n\#}\eta_n$ are almost optimal with respect to $C_{\Sigma_n}$ and their supports are bounded in $\Theta$ by $L + \varepsilon_n$. □

Our next aim is to study the properties of the optimal measures $\mu$: in particular, we want to determine whether or not a relaxed solution $\mu$ corresponds to a classical solution $\Sigma$.

**Definition 4.7.** A measure $\mu$ optimal for the relaxed functional $\mathfrak{F}$ is called *minimal* if any measure $\nu \leq \mu$, $\nu \neq \mu$ is not optimal for $\mathfrak{F}$.

*Remark 4.8.* Notice that, thanks to the lower semicontinuity and the coercivity of $\mathfrak{F}$, proven in Proposition 4.5, the existence of an optimal measure for $\mathfrak{F}$ is straightforward. In particular, the set of these optimal measures is a non-empty, bounded and weakly* closed subset of $\mathcal{M}^+(\Omega)$; therefore, since the map $\mu \mapsto \|\mu\|$ is weakly* l.s.c., among the optimal measures $\mu$ there are those minimizing the norm $\|\mu\|$. Finally, an immediate application of Zorn's Lemma gives us the existence of minimal optimal measures; more precisely, for any optimal measure $\mu$ there exists a minimal optimal measure $\mu' \leq \mu$.

A first property that we are now able to show is that any minimal optimal measure is absolutely continuous with respect to $\mathscr{H}^1$ –in fact, in most of the cases *all* the optimal measures are minimal, see Proposition 4.20.

**Lemma 4.9.** *Each minimal optimal measure is absolutely continuous with respect to* $\mathscr{H}^1$.

*Proof.* We argue by contradiction assuming the existence of a set $B$ such that $\mathscr{H}^1(B) = 0$ but $\mu(B) > 0$. Take a sequence $\{\Sigma_n\}$ of Borel sets such that

$$\mathscr{H}^1 \llcorner \Sigma_n =: \mu_n \overset{*}{\rightharpoonup} \mu$$

and $\mathfrak{F}(\Sigma_n) \to \mathfrak{F}(\mu)$, which is possible in view of the definition (4.1) of $\mathfrak{F}$. Fixed now $\varepsilon > 0$, by definition of the Hausdorff measure we can take countably many open balls $B_i$ with radii $r_i$, such that $S_\varepsilon := \cup B_i$ contains $B$ and $\sum r_i < \varepsilon$. Define then

$$\Sigma_{n,\varepsilon} := \Sigma_n \setminus S_\varepsilon , \qquad \mu_{n,\varepsilon} := \mathscr{H}^1 \llcorner \Sigma_{n,\varepsilon} .$$

Choose now an arbitrary path $\theta \in \Theta$: we claim the existence of a path $\alpha(\theta)$ having the same endpoints of $\theta$ and such that

$$\alpha(\theta) \setminus S_\varepsilon \subseteq \theta \setminus S_\varepsilon , \qquad \mathscr{H}^1(\alpha(\theta) \cap S_\varepsilon) \leq 2\varepsilon .$$

To show this fact, define $\theta_1 := \theta$ if $\theta \cap B_1 = \emptyset$; otherwise, fixed an arbitrary parametrization of $\theta$, let $t_1$ and $t_2$ be the first and the last instant such that $\theta(t) \in \overline{B_1}$, and define $\theta_1$ to be the path that equals $\theta$ in $[0, t_1] \cup [t_2, 1]$, and that is the line segment connecting $\theta(t_1)$ to $\theta(t_2)$ in $[t_1, t_2]$. In this way,

$$\theta_1 \setminus B_1 \subseteq \theta \setminus B_1 , \qquad \mathscr{H}^1(\theta_1 \cap B_1) \leq 2r_1 .$$

Note that the map $\theta \mapsto \theta_1$ is Borel since it is continuous on each of the sets

$$\{\theta \in \Theta : \theta \cap \overline{B_1} \neq \emptyset\}, \qquad \{\theta \in \Theta : \theta \cap \overline{B_1} = \emptyset\}.$$

In the same way, replacing $\theta$ by $\theta_1$ and $B_1$ by $B_2$, we define $\theta_2$ so that

$$\theta_2 \setminus B_2 \subseteq \theta_1 \setminus B_2, \qquad \mathscr{H}^1(\theta_2 \cap B_2) \leq 2r_2.$$

Analogously, the map $\theta \mapsto \theta_2$ is continuous on each of the four sets

$$\begin{aligned} &\{\theta \in \Theta : \theta \cap \overline{B_1} \neq \emptyset, \theta \cap \overline{B_2} \neq \emptyset\}, \quad \{\theta \in \Theta : \theta \cap \overline{B_1} = \emptyset, \theta \cap \overline{B_2} \neq \emptyset\}, \\ &\{\theta \in \Theta : \theta \cap \overline{B_1} = \emptyset, \theta \cap \overline{B_2} \neq \emptyset\}, \quad \{\theta \in \Theta : \theta \cap \overline{B_1} = \emptyset, \theta \cap \overline{B_2} = \emptyset\}, \end{aligned}$$

and hence is Borel. Iterating this procedure, we find a sequence $\{\theta_n\}$ of paths connecting $\theta(0)$ and $\theta(1)$ and, by construction and recalling that $\sum_{i\in\mathbb{N}} r_i \leq \varepsilon$, we deduce that the paths $\theta_n$ uniformly converge to a path $\alpha(\theta)$ which equals $\theta$ outside of $S_\varepsilon$ and with

$$\mathscr{H}^1(\alpha(\theta) \cap S_\varepsilon) \leq \sum_i 2r_i \leq 2\varepsilon,$$

so that our claim is proved. Moreover, each map $\theta \mapsto \theta_n$ is Borel and hence so is their pointwise limit $\alpha : \Theta \to \Theta$ (in fact, $\Theta$ is the countable union of Borel sets on each of which the map $\alpha$ is continuous).

Applying now Lemma 4.6, we find a t.p.m. $\eta_n$ whose support is bounded in $\Theta$ by $L$ and which is optimal with respect to $\overline{C}_{\Sigma_n}$. Since $A$ is uniformly continuous on $[0, L]$, we denote by $\omega$ its modulus of continuity. Then, for every $\theta \in \operatorname{spt} \eta_n$ one has by construction

$$\mathscr{H}^1(\alpha(\theta) \cap \Sigma_{n,\varepsilon}) \leq \mathscr{H}^1(\theta \cap \Sigma_n)$$

and

$$\mathscr{H}^1(\alpha(\theta) \setminus \Sigma_{n,\varepsilon}) \leq \mathscr{H}^1(\theta \setminus \Sigma_n) + 2\varepsilon,$$

so that

$$\bar{\delta}_{\Sigma_{n,\varepsilon}}(\alpha(\theta)) \leq \bar{\delta}_{\Sigma_n}(\theta) + \omega(2\varepsilon);$$

as a consequence,

$$\begin{aligned} \mathfrak{F}(\mu_{n,\varepsilon}) &\leq \overline{C}_{\Sigma_{n,\varepsilon}}(\alpha_\# \eta_n) + H(\|\mu_{n,\varepsilon}\|) \leq \overline{C}_{\Sigma_n}(\eta_n) + \omega(2\varepsilon) + H(\|\mu_n\|) \\ &= \mathfrak{F}(\mu_n) + \omega(2\varepsilon), \end{aligned} \tag{4.3}$$

since $\mu_{n,\varepsilon} \leq \mu_n$ by definition and $H$ is nondecreasing. Since $\mu_n \xrightarrow{*} \mu$ and $\mu_{n,\varepsilon} \leq \mu_n$, up to a subsequence we can assume that $\mu_{n,\varepsilon} \xrightarrow{*} \mu_\varepsilon$ with $\mu_\varepsilon \leq \mu$. In particular,

$$\mu_\varepsilon(B) \leq \mu_\varepsilon(S_\varepsilon) = 0,$$

since $S_\varepsilon$ is open and $\mu_{n,\varepsilon}(S_\varepsilon) = 0$ for any $n \in \mathbb{N}$.

We select now a sequence $\varepsilon_j \searrow 0$, finding the measures $\mu_{\varepsilon_j}$ as shown above; we can choose the balls $B_h$ related to each $\varepsilon_j$ in such a way that $S_{\varepsilon_i} \subseteq S_{\varepsilon_j}$ whenever $i \geq j$: as a consequence, we have

$$\mu_{n,\varepsilon_i} \geq \mu_{n,\varepsilon_j} \qquad \text{whenever} \qquad i \geq j\,.$$

Hence, passing to a weak* limit as $n \to \infty$ (choosing a common subsequence of indices), we have $\mu_{\varepsilon_i} \geq \mu_{\varepsilon_j}$ for $i \geq j$. Therefore, $j \mapsto \mu_{\varepsilon_j}$ is an increasing sequence of measures bounded by $\mu$ and with the property that $\mu_{\varepsilon_j}(B) = 0$ for any $j$. We derive that $\mu_{\varepsilon_j}$ converge strongly as $j \to \infty$ to a measure $\bar{\mu}$ with

$$\bar{\mu} \leq \mu\,, \qquad \bar{\mu}(B) = 0\,, \qquad \mathfrak{F}(\bar{\mu}) \leq \mathfrak{F}(\mu)\,;$$

the latter property immediately follows by (4.3) recalling the lower semicontinuity of $\mathfrak{F}$, the fact that $\mathfrak{F}(\mu_n) \to \mathfrak{F}(\mu)$, and passing to the limit in (4.3) first as $n \to \infty$, and then as $\varepsilon \searrow 0$.

The measure $\bar{\mu}$ is then optimal because so is $\mu$; but

$$\bar{\mu}(B) = 0 < \mu(B)\,,$$

hence $\bar{\mu} \neq \mu$; this contradicts the fact that $\mu$ is a minimal optimal measure. □

We are able now to prove a stronger result, namely, that the minimal optimal measures for $\mathfrak{F}$ are concentrated on one-dimensional sets.

**Lemma 4.10.** *Every minimal optimal measure $\mu$ for $\mathfrak{F}$ has the form*

$$\mu = \varphi \mathscr{H}^1 \llcorner \Sigma$$

*for a suitable Borel set $\Sigma$ and a Borel function $\varphi : \Sigma \to \mathbb{R}^+$.*

*Proof.* Let $\mu$ be a minimal optimal measure, so that thanks to Lemma 4.9 we have $\mu \ll \mathscr{H}^1$. Thanks to Theorem 3.2 in [64], the thesis is achieved if

$$\theta_1^*(\mu, x) := \limsup_{\varepsilon \to 0} \frac{\mu\big(B(x,\varepsilon)\big)}{\varepsilon} > 0 \qquad \text{for } \mu\text{–a.e. } x\,. \tag{4.4}$$

We prove now (4.4) by contradiction; to this aim, take a sequence

$$\mu_n := \mathscr{H}^1 \llcorner \Sigma_n$$

such that

$$\mu_n \stackrel{*}{\rightharpoonup} \mu\,, \qquad \mathfrak{F}(\mu) = \lim \mathfrak{F}(\Sigma_n)\,.$$

If (4.4) does not hold, there is a constant $C$ such that $\mu(X) > C > 0$, where

$$X := \{x : \theta_1^*(\mu, x) = 0\}.$$

Fixed now a small $\delta > 0$ (that we will eventually send to 0), we know that $\mu(X^\delta) > C$, defining

$$X^\delta := \{x : \theta_1^*(\mu, x) < \delta\}.$$

We deduce that $\mu(X_\varepsilon^\delta) > C$ for $\varepsilon$ small enough (depending on $\delta$), where

$$X_\varepsilon^\delta := \{x : \mu(\overline{B}(x,\varepsilon)) < \delta\varepsilon\}.$$

Note also that, since we used closed balls $\overline{B}(x,\varepsilon)$, the set $X_\varepsilon^\delta$ is open.

Keeping fixed $\delta$ and $\varepsilon$, for $n$ large enough we define

$$\widetilde{\Sigma}_n := \Sigma_n \setminus X_\varepsilon^\delta$$

and $\tilde{\mu}_n := \mathscr{H}^1 \llcorner \widetilde{\Sigma}_n$. We aim to show the existence of some $K(\delta) \xrightarrow[\delta \to 0]{} 0$ such that

$$\mathfrak{F}(\tilde{\mu}_n) \leq \mathfrak{F}(\mu_n) + K(\delta). \tag{4.5}$$

We claim that (4.5) implies the thesis. Indeed, let $\mu_\delta$ be a weak* limit of a subsequence of $\{\tilde{\mu}_n\}$: since $\tilde{\mu}_n \leq \mu_n$ and $\mu_n \overset{*}{\rightharpoonup} \mu$, one has $\mu_\delta \leq \mu$; moreover, since $\tilde{\mu}_n(X_\varepsilon^\delta) = 0$ and $X_\varepsilon^\delta$ is open, we obtain $\mu_\delta(X_\varepsilon^\delta) = 0$ while $\mu(X_\varepsilon^\delta) > C$, so that

$$\|\mu_\delta\| \leq \|\mu\| - C.$$

As a consequence, a weak* limit $\bar{\mu}$ of a subsequence of $\{\mu_\delta\}$ as $\delta \searrow 0$ satisfies $\bar{\mu} \leq \mu$ and $\|\bar{\mu}\| \leq \|\mu\| - C$, hence $\bar{\mu} \neq \mu$. But if (4.5) is true, as in Lemma 4.9 $\mathfrak{F}(\bar{\mu}) \leq \mathfrak{F}(\mu)$ and this contradicts the fact that $\mu$ is a minimal optimal measure.

In order to get (4.5), since $\mathscr{H}^1(\widetilde{\Sigma}_n) \leq \mathscr{H}^1(\Sigma_n)$, recalling (2.11) and (2.22), it suffices to show that for $n$ large enough (depending on $\delta$) one has

$$d_{\widetilde{\Sigma}_n}(x,y) \leq d_{\Sigma_n}(x,y) + K(\delta) \qquad \forall\,(x,y) \in \Omega \times \Omega. \tag{4.6}$$

By Lemma 2.16, we get that the functions $d_{\widetilde{\Sigma}_n}$ and $d_{\Sigma_n}$ are equi-uniformly continuous over $\Omega \times \Omega$, since their moduli of continuity can be estimated by means of the modulus of continuity of $A$ in $[0, L]$. Therefore, by the boundedness of $\Omega$, it suffices to verify the inequality (4.6) for a finite number of pairs $(x_i, y_i)$, and hence for a given pair $(x, y)$.

Let now $(x,y) \in \Omega \times \Omega$ be fixed and choose paths $\theta_n$ almost optimal between $x$ and $y$ with respect to $\Sigma_n$ in the sense that

$$\delta_{\Sigma_n}(\theta_n) \leq d_{\Sigma_n}(x,y) + \frac{1}{n}.$$

It suffices to show that

$$\delta_{\widetilde{\Sigma}_n}(\theta_n) \le \delta_{\Sigma_n}(\theta_n) + K(\delta) \tag{4.7}$$

for all $n$ large enough. To show (4.7), we consider $\theta_n$ as parametrized by constant speed and we recall that by Lemma 2.16 one can assume $\mathscr{H}^1(\theta_n) \le L$ for $n$ large enough; hence, up to a subsequence, the paths $\theta_n$ uniformly converge to some path $\theta$. We claim that there exists an $\bar{n} = \bar{n}(\delta, x, y)$ such that for every $t \in [0,1]$ and $n \ge \bar{n}$ one has

$$\mu_n\big(\overline{B}(\theta_n(t), \varepsilon/2)\big) \le \mu\big(\overline{B}(\theta_n(t), \varepsilon)\big) + \delta\varepsilon\,. \tag{4.8}$$

In fact, for each fixed $\bar{t} \in [0,1]$ we have

$$\mu_n\Big(\overline{B}\Big(\theta_n(\bar{t}), \frac{4}{6}\varepsilon\Big)\Big) \le \mu\Big(\overline{B}\Big(\theta_n(\bar{t}), \frac{5}{6}\varepsilon\Big)\Big) + \delta\varepsilon\,, \tag{4.9}$$

whenever $n \ge \bar{n}(\bar{t}, \delta, x, y)$. Since the paths $\theta_n$ are parametrized by constant speed and have length bounded by $L$, then

$$\overline{B}\Big(\theta_n(t), \frac{3}{6}\varepsilon\Big) \subseteq \overline{B}\Big(\theta_n(\bar{t}), \frac{4}{6}\varepsilon\Big), \quad \overline{B}\Big(\theta_n(\bar{t}), \frac{5}{6}\varepsilon\Big) \subseteq \overline{B}\Big(\theta_n(t), \varepsilon\Big), \tag{4.10}$$

whenever

$$|t - \bar{t}| \le \frac{\varepsilon}{6L}\,.$$

From (4.9) and (4.10) we deduce the validity of (4.8) for all $n \ge \bar{n}(\bar{t}, \delta, x, y)$ and for all $t$ satisfying

$$|t - \bar{t}| \le \frac{\varepsilon}{6L}\,.$$

Hence, by compactness of $[0,1]$, we get the validity of (4.8) for all $n \ge \bar{n}(\delta, x, y)$ (i.e. with $\bar{n}$ independent of $t$).

Take now $n \ge \bar{n}(\delta, x, y)$ and $t \in [0,1]$ such that $\theta_n(t) \in X_\varepsilon^\delta$: the definition of $X_\varepsilon^\delta$ and the above estimate ensure that

$$\mu_n\big(\overline{B}(\theta_n(t), \varepsilon/2)\big) \le 2\delta\varepsilon\,.$$

Setting then

$$D_t := [t - \varepsilon/(2v_n), t + \varepsilon/(2v_n)]\,,$$

where

$$v_n = \mathscr{H}^1(\theta_n) = |\theta_n'(t)|$$

for every $t \in [0,1]$, one has that $\mathscr{H}^1\big(\theta_n(D_t)\big) = \varepsilon$, while

$$\mathscr{H}^1 \llcorner \Sigma_n\big(\theta_n(D_t)\big) \le \mu_n\big(\overline{B}(\theta_n(t), \varepsilon/2)\big) \le 2\delta\varepsilon = 2\delta\mathscr{H}^1\big(\theta_n(D_t)\big)\,.$$

Now we can cover $\theta_n^{-1}\big(X_\varepsilon^\delta\big)$ by a finite number of intervals of the form of $D_t$ in such a way that no instant $t \in [0,1]$ belongs to more than two such intervals. As a consequence, we immediately infer

$$\mathscr{H}^1(\theta_n) \geq \frac{\mathscr{H}^1 \llcorner \Sigma_n(\theta_n \cap X^\delta_\varepsilon)}{4\delta} \tag{4.11}$$

so, defining

$$a := \mathscr{H}^1(\theta_n \setminus \Sigma_n), \quad b := \mathscr{H}^1 \llcorner \Sigma_n(\theta_n \cap X^\delta_\varepsilon), \quad c := \mathscr{H}^1 \llcorner \widetilde{\Sigma}_n(\theta_n),$$

we may summarize what obtained by the equalities

$$\delta_{\Sigma_n}(\theta_n) = A(a) + B(b+c), \qquad \delta_{\widetilde{\Sigma}_n}(\theta_n) = A(a+b) + B(c); \tag{4.12}$$

moreover,

$$a + b + c = v_n \leq L.$$

Finally, (4.11) and the definition of $a$, $b$ and $c$ give

$$a + b + c \geq \frac{b}{4\delta}.$$

This implies that $b \leq 4\delta\, L$ and hence, by (4.12) and by the uniform continuity of $A$ on the compact set $[0, L] \subseteq \mathbb{R}$, we get

$$\begin{aligned} \delta_{\widetilde{\Sigma}_n}(\theta_n) - \delta_{\Sigma_n}(\theta_n) &= A(a+b) + B(c) - A(a) - B(b+c) \\ &\leq A(a+b) - A(a) \leq \omega_A(4\delta L), \end{aligned}$$

where $\omega_A$ stands for the modulus of continuity of $A$ over the interval $[0, L]$. Hence, (4.7) is proved with

$$K(\delta) := \omega_A(4\delta L),$$

which therefore concludes the proof. □

We can go now further, showing that the set $\Sigma$ in last lemma can be chosen rectifiable.

**Lemma 4.11.** *Any minimal optimal measure is concentrated on a rectifiable set $\Sigma$.*

*Proof.* Let us take an arbitrary optimal measure $\beta$ and a minimizing sequence $\beta_n = \mathscr{H}^1 \llcorner \Sigma_n$ for the functional $\mathfrak{F}$, i.e. $\beta_n \xrightarrow{*} \beta$ and $\mathfrak{F}(\beta_n) \to \mathfrak{F}(\beta)$. The strategy to achieve the conclusion will be to construct an optimal measure $\mu \leq \beta$ concentrated on a rectifiable set: by Definition 4.7 of minimality of an optimal measure, this yields the thesis.

If $\mathfrak{F}(\beta) = \mathfrak{F}(\emptyset)$, there is nothing to prove. Otherwise,

$$K := \mathfrak{F}(\emptyset) - \mathfrak{F}(\beta) > 0,$$

so that for $n$ sufficiently large one has

$$\mathfrak{F}(\emptyset) - \mathfrak{F}(\beta_n) > \frac{K}{2}\,.$$

According to Proposition 2.14, for $n$ large enough we take a t.p.m. $\eta_n$ such that

$$C_{\Sigma_n}(\eta_n) - MK(\Sigma_n) < \mathfrak{F}(\emptyset) - \mathfrak{F}(\beta_n) - K/2\,.$$

Therefore,

$$\begin{aligned}
&\int_\Theta A(\mathscr{H}^1(\theta)) - A(\mathscr{H}^1(\theta \setminus \Sigma_n))\, d\eta_n \geq \\
&\quad \geq \int_\Theta A(\mathscr{H}^1(\theta)) - \Big(A(\mathscr{H}^1(\theta \setminus \Sigma_n)) + B(\mathscr{H}^1(\theta \cap \Sigma_n))\Big)\, d\eta_n \\
&\quad = C_\emptyset(\eta_n) - C_{\Sigma_n}(\eta_n) \\
&\quad \geq \mathfrak{F}(\emptyset) - C_{\Sigma_n}(\eta_n) > \mathfrak{F}(\beta_n) - MK(\Sigma_n) + K/2 \\
&\quad = \mathfrak{F}(\Sigma_n) - MK(\Sigma_n) + K/2 \geq K/2\,.
\end{aligned} \tag{4.13}$$

Recalling that all the t.p.m.'s have unitary total mass, since

$$\|f^+\| = \|f^-\| = 1\,,$$

we deduce the existence of paths $\theta_n \in \operatorname{spt} \eta_n$ such that

$$A(\mathscr{H}^1(\theta_n)) - A(\mathscr{H}^1(\theta_n \setminus \Sigma_n)) \geq \frac{K}{2}\,. \tag{4.14}$$

By Lemma 4.6, since $\theta_n \in \operatorname{spt} \eta_n$ and the sequence $\{\eta_n\}$ is almost optimal with respect to $C_{\Sigma_n}$, we may assume that

$$\mathscr{H}^1(\theta_n) \leq L\,;$$

thus, thanks to the uniform continuity of $A$ in $[0, L]$, from (4.14) we infer that

$$\begin{aligned}
\beta_n(\theta_n) &= \mathscr{H}^1 \llcorner \Sigma_n(\theta_n) = \mathscr{H}^1\big(\theta_n \cap \Sigma_n\big) = \mathscr{H}^1(\theta_n) - \mathscr{H}^1(\theta_n \setminus \Sigma_n) \\
&\geq \omega^{-1}(K/2)\,,
\end{aligned}$$

where $\omega$ stands for the modulus of continuity of $A$ in $[0, L]$, and $\omega^{-1}(K/2)$ stands for the biggest number $x > 0$ such that $\omega(x) \leq K/2$ (this is a slight abuse of notations, since $\omega$ may be not *strictly* increasing). Since $\{\theta_n\}$ is a sequence of paths of Euclidean length bounded by $L$ we may assume, up to a subsequence, that the paths $\theta_n$ uniformly converge to some path $\theta \in \Theta$, thus in particular the respective traces $\theta_n([0,1])$ converge in the Hausdorff distance to $C_1 = \theta([0,1])$. Again up to a subsequence, we assume that

$$\beta_n \llcorner \theta_n \stackrel{*}{\rightharpoonup} \mu_1$$

so that, since $C_1$ is closed, we deduce that $\mu_1$ is concentrated in $C_1$: in fact, for every open set $U$ such that dist $(U, C_1) > 0$ one has

$$\mu_1(U) \le \liminf_{n\to\infty} \beta_n \llcorner \theta_n(U) = 0 .$$

Thanks to Lemma 4.1, we have

$$\|\mu_1\| \ge \omega^{-1}(K/2) .$$

Noticing that $\mu_1 \le \beta$, we obtain the estimates

$$\beta(C_1) \ge \mu_1(C_1) \ge \omega^{-1}(K/2) > 0 .$$

We want now to iterate the above argument: to this aim, we call

$$\widetilde{\Sigma}_n := \Sigma_n \setminus \theta_n$$

so that

$$\tilde{\beta}_n = \mathcal{H}^1 \llcorner \widetilde{\Sigma}_n \stackrel{*}{\rightharpoonup} \beta - \mu_1 .$$

We use now an argument similar to the previous one: if $\mathfrak{F}(\mu_1) \le \mathfrak{F}(\beta)$, then $\mu_1$ is an optimal measure and we conclude. Otherwise,

$$K' := \mathfrak{F}(\mu_1) - \mathfrak{F}(\beta) > 0 ,$$

so that for $n$ sufficiently large

$$\rho_n := \mathfrak{F}(\beta_n \llcorner \theta_n) - \mathfrak{F}(\beta_n) - \frac{K'}{2} > 0 :$$

indeed, we know that $\mathfrak{F}(\beta_n) \to \mathfrak{F}(\beta)$ and by lower semicontinuity of $\mathfrak{F}$ one has

$$\liminf \mathfrak{F}(\beta_n \llcorner \theta_n) \ge \mathfrak{F}(\mu_1) .$$

Arguing as in (4.13), choosing the sequence $\{\eta_n\}$ almost optimal for $C_{\Sigma_n}$ ($\eta_n$ is possibly different from the ones of the first step), noticing that

$$\beta_n \llcorner \theta_n = \mathcal{H}^1 \llcorner (\Sigma_n \cap \theta_n)$$

and recalling that

$$\mathcal{H}^1(\Sigma_n) \ge \mathcal{H}^1(\Sigma_n \cap \theta_n)$$

for any $n$, one has the relationship

$$\begin{aligned} C_{\Sigma_n \cap \theta_n}(\eta_n) - C_{\Sigma_n}(\eta_n) &\ge MK(\Sigma_n \cap \theta_n) - MK(\Sigma_n) - \rho_n \\ &= \mathfrak{F}(\Sigma_n \cap \theta_n) - H(\mathcal{H}^1(\Sigma_n \cap \theta_n)) - \mathfrak{F}(\Sigma_n) + H(\mathcal{H}^1(\Sigma_n)) - \rho_n \\ &\ge \mathfrak{F}(\Sigma_n \cap \theta_n) - \mathfrak{F}(\Sigma_n) - \rho_n = K'/2 . \end{aligned}$$

Therefore, for $n$ large enough, one has

$$\begin{aligned}
&\int_\Theta A\big(\mathscr{H}^1(\theta\setminus\Sigma_n)+\mathscr{H}^1(\theta\cap\Sigma_n\setminus\theta_n)\big)-A\big(\mathscr{H}^1(\theta\setminus\Sigma_n)\big)\,d\eta_n\\
&\quad\geq\int_\Theta\Big(A\big(\mathscr{H}^1(\theta\setminus\Sigma_n)+\mathscr{H}^1(\theta\cap\Sigma_n\setminus\theta_n)\big)\\
&\qquad+B\big(\mathscr{H}^1(\theta\cap\Sigma_n)-\mathscr{H}^1(\theta\cap\Sigma_n\setminus\theta_n)\big)\Big)\\
&\qquad-\Big(A\big(\mathscr{H}^1(\theta\setminus\Sigma_n)\big)+B\big(\mathscr{H}^1(\theta\cap\Sigma_n)\big)\Big)\,d\eta_n\\
&\quad=\int_\Theta\Big(A\big(\mathscr{H}^1(\theta\setminus(\Sigma_n\cap\theta_n))\big)+B\big(\mathscr{H}^1(\theta\cap(\Sigma_n\cap\theta_n))\big)\Big)-\delta_{\Sigma_n}(\theta)\,d\eta_n\\
&\quad=C_{\Sigma_n\cap\theta_n}(\eta_n)-C_{\Sigma_n}(\eta_n)\geq\frac{K'}{2}.
\end{aligned}$$

As before, we deduce the existence of a sequence of paths $\{\sigma_n\}\in\operatorname{spt}\eta_n$ of lengths bounded by $L$ and with

$$A\big(\mathscr{H}^1(\sigma_n\setminus\Sigma_n)+\mathscr{H}^1(\sigma_n\cap\Sigma_n\setminus\theta_n)\big)-A\big(\mathscr{H}^1(\sigma_n\setminus\Sigma_n)\big)\geq\frac{K'}{2},$$

so that

$$\tilde\beta_n(\sigma_n)=\mathscr{H}^1(\sigma_n\cap\Sigma_n\setminus\theta_n)\geq\omega^{-1}(K'/2).$$

Again, by Lemma 4.1 we find a measure $\mu_2$ concentrated in a curve $C_2\in\Theta$ of length bounded by $L$ ($C_2$ is the Hausdorff limit of the traces $\sigma_n([0,1])$), such that

$$\tilde\beta_n\llcorner\sigma_n\overset{*}{\rightharpoonup}\mu_2,\qquad\text{and}\qquad\|\mu_2\|\geq\omega^{-1}(K'/2).$$

Recalling that $\tilde\beta_n\overset{*}{\rightharpoonup}\beta-\mu_1$, we deduce $\mu_1+\mu_2\leq\beta$.

We iterate this argument, finding measures $\mu_i$ concentrated in curves $C_i$ and such that

$$\hat\mu_n:=\mu_1+\mu_2+\cdots+\mu_n$$

is an increasing sequence of positive measures with $\hat\mu_n\leq\beta$; since $\|\beta\|<+\infty$, we deduce that $\hat\mu_n$ converges strongly to a measure $\mu_\infty$ with $\mu_\infty\leq\beta$. Now, by construction we have that

$$\big\|\hat\mu_{n+1}-\hat\mu_n\big\|=\|\mu_{n+1}\|\geq\omega^{-1}\Big(\big(\mathfrak{F}(\hat\mu_n)-\mathfrak{F}(\beta)\big)/2\Big);$$

but since $\hat\mu_n\to\mu_\infty$ strongly, we infer that

$$\big\|\hat\mu_{n+1}-\hat\mu_n\big\|\longrightarrow0,$$

then

$$\omega^{-1}\Big(\big(\mathfrak{F}(\hat{\mu}_n) - \mathfrak{F}(\beta)\big)/2\Big) \longrightarrow 0\,.$$

Keeping in mind that $\omega$ is the modulus of continuity of the function $A$ in $[0, L]$, we infer that $\mathfrak{F}(\hat{\mu}_n) \to \mathfrak{F}(\beta)$ so that, by lower semicontinuity of $\mathfrak{F}$, the measure $\mu_\infty$ is an optimal measure less than $\beta$, as we were looking for. □

Summarizing, we know that there exists an optimal measure $\mu$ concentrated in a 1−rectifiable set $\Sigma$, and absolutely continuous with respect to $\mathscr{H}^1$. It follows that $\mu = \varphi\mathscr{H}^1 \llcorner \Sigma$, so we found a relaxed solution which is, in a certain sense, similar to the classical one: if $\varphi \equiv 1$, we have indeed a classical solution. We will prove in a moment that $\varphi \le 1$: this is quite reasonable, since we can imagine that covering a path with $\varphi > 1$ has a cost strictly greater than covering the same path with $\varphi \equiv 1$: even though this argument is far from being formal, this is more or less the idea of the proof of Theorem 4.14. On the other hand, the case $\varphi < 1$ is not meaningless: in fact, following a path of length $l$ on a network where $\varphi \equiv p \in [0, 1]$ can be interpreted as covering a length $pl$ by train and the remaining $(1 - p)l$ by own means, which may be, in some cases, better than covering the whole length $l$ by train. More precisely, one can imagine that in some situation this is indeed the best case, which is exactly what we will prove in the example of Section 4.3. This intuitive discussion will become clearer in light of Proposition 4.15.

In the sequel, we will use for simplicity the following notation.

**Definition 4.12.** We denote by $\mathcal{M}_1^+(\Omega)$ the set of those measures $\mu \in \mathcal{M}^+(\Omega)$ which can be represented as

$$\mu = \varphi\mathscr{H}^1 \llcorner \Sigma\,,$$

where $\Sigma$ is the union of countably many Lipschitz paths of uniformly bounded length, and

$$\varphi : \Sigma \to [0, 1]$$

is a Borel function.

**Definition 4.13.** Given a path $\theta \in \Theta$, and given two points of $\theta$, namely $P = \theta(s)$ and $Q = \theta(t)$ with $t \ge s$, we will denote by $\widetilde{PQ}$ the path $\theta \llcorner [s, t] \in \Theta$; we will use this notation only when the particular path $\theta$ containing $P$ and $Q$ will be clear from the context.

We show now, as anticipated, the existence of an optimal measure $\mu$ with $\varphi \le 1$. In view of Definition 4.12, we can equivalently say that there exists an optimal measure $\mu \in \mathcal{M}_1^+(\Omega)$.

**Theorem 4.14.** *There is an optimal measure $\mu \in \mathcal{M}_1^+(\Omega)$; more precisely, each optimal measure $\mu$ which is minimal in the sense of Definition 4.7 belongs to $\mathcal{M}_1^+(\Omega)$.*

*Proof.* Let $\mu$ be a minimal optimal measure, which is concentrated in a rectifiable set

$$\Sigma = \cup_{i\in\mathbb{N}}\theta_i$$

where the paths $\theta_i \in \Theta$ have lengths bounded by a constant $L$ by Lemmas 4.6 and 4.11. Let $\{\Sigma_n\}$ be a minimizing sequence in the definition of $\mathfrak{F}(\mu)$, that is,

$$\mu_n := \mathscr{H}^1 \llcorner \Sigma_n \stackrel{*}{\rightharpoonup} \mu$$

and $\mathfrak{F}(\mu_n) \to \mathfrak{F}(\mu)$. We claim that $\mu \in \mathcal{M}_1^+(\Omega)$: if it were false, there would be an index $i \in \mathbb{N}$ and a Lipschitz path

$$\Delta_0 = \widetilde{PQ} \subseteq \theta_i \subseteq \Sigma$$

such that $\mu(\Delta_0) > \mathscr{H}^1(\Delta_0)$.

In order to find a competitor to $\mu$, that will lead to a contradiction, we fix an $\varepsilon > 0$ (that we will eventually lead to 0) and we let $C_\varepsilon$ to stand for the open $\varepsilon-$tubular neighborhood of $\Delta_0$, that is the set of those points of $\Omega$ the minimal distance of which from $\theta_i$ is strictly less than $\varepsilon$ and is reached at some point of $\Delta_0$. Notice that, up to an arbitrary small movement of $P$ and $Q$, $\Delta_0$ can be chosen in such a way that, for all $\varepsilon > 0$ except for countably many,

$$\mu\big(\partial C_\varepsilon\big) = 0\,, \qquad \text{and} \qquad \forall n\in\mathbb{N}\,,\ \mu_n\big(\partial C_\varepsilon\big) = 0\,. \tag{4.15}$$

In the rest of the proof, we deal only with numbers $\varepsilon > 0$ for which (4.15) holds.

We now set, for any $n$,

$$\Sigma_{n,\varepsilon} := \big(\Sigma_n \setminus C_\varepsilon\big) \cup \Delta_0, \qquad \mu_{n,\varepsilon} := \mathscr{H}^1 \llcorner \Sigma_{n,\varepsilon}\,. \tag{4.16}$$

In short, from every $\Sigma_n$ we drop the whole tubular neighborhood $C_\varepsilon$, and then we add the curve $\Delta_0$.

We want now to check that $\mathfrak{F}\big(\Sigma_{n,\varepsilon}\big)$ is not much larger than $\mathfrak{F}\big(\Sigma_n\big)$. To this aim, we fix a $\delta < \varepsilon^2$ and we define $D_\delta \subseteq C_\varepsilon$ the $\delta$-tubular neighborhood of the part of $\Delta_0$ having distance more than $\varepsilon$ from both $P$ and $Q$; formally,

$$D_\delta := \Big\{x \in \Omega : \exists y \in \Delta_0,\ \mathscr{H}^1(\widetilde{Py}) > \varepsilon, \mathscr{H}^1(\widetilde{yQ}) > \varepsilon,$$
$$\operatorname{dist}(x,\Delta_0) = |y-x| < \delta\Big\}$$

(see Figure 4.1). Notice that, since $\mu$ is a finite measure, we have

$$\mu(C_\varepsilon) = \mu(\Delta_0) + K_1(\varepsilon)$$

with $K_1(\varepsilon) \xrightarrow[\varepsilon\to 0]{} 0$ and, exactly as in (4.15), for all but countably many $\delta$ one has

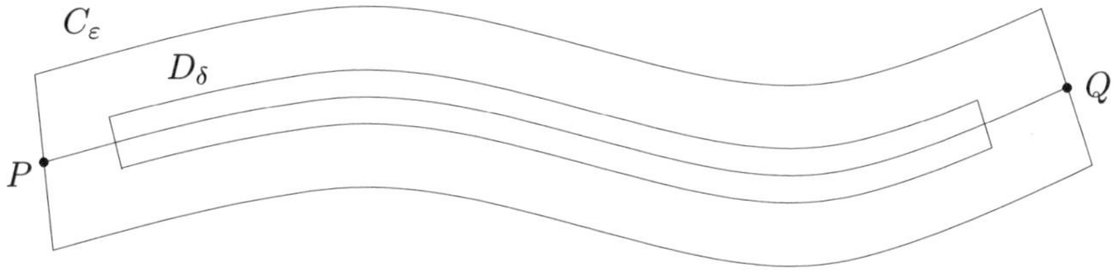


**Fig. 4.1** Construction of the proof of Theorem 4.14

$$\mu(\partial D_\varepsilon) = 0, \qquad \text{and} \qquad \forall n \in \mathbb{N},\ \mu_n(\partial D_\varepsilon) = 0\,. \tag{4.17}$$

Moreover,

$$\Sigma_\varepsilon := \cap_{\delta>0} D_\delta$$

is a Lipschitz path (since it is contained in $\Delta_0$), namely the part of $\Delta_0$ made by those points having distance more than $\varepsilon$ from both $P$ and $Q$: therefore, for any $\delta > 0$ one has $\mu(D_\delta) \geq \mu(\Sigma_\varepsilon)$. Since

$$\mu(\Sigma_\varepsilon) \nearrow \mu(\Delta_0) \qquad \text{as} \qquad \varepsilon \searrow 0\,,$$

it follows that

$$\mu(C_\varepsilon \setminus D_\delta) \leq \mu(C_\varepsilon) - \mu(\Sigma_\varepsilon) = K_1(\varepsilon) + \mu(\Delta_0) - \mu(\Sigma_\varepsilon) =: K_2(\varepsilon)\,,$$

where again $K_2(\varepsilon) \underset{\varepsilon\to 0}{\longrightarrow} 0$. Recalling that $\mu_n \overset{*}{\rightharpoonup} \mu$, (4.15) and (4.17), we derive that

$$\mu_n(C_\varepsilon \setminus D_\delta) \to \mu(C_\varepsilon \setminus D_\delta)$$

as $n \to \infty$, and hence, for $n$ large enough, one has

$$\mu_n(C_\varepsilon \setminus D_\delta) \leq 2K_2(\varepsilon)\,. \tag{4.18}$$

Using the projection map from the tubular neighborhood $C_\varepsilon$ onto $\Delta_0$, which is defined for all sufficiently small $\varepsilon$, we may define a Lipschitz map $\alpha_\delta : \Omega \to \Omega$ such that (recall that $\delta < \varepsilon^2$)

**i)** $\alpha_\delta \equiv \mathrm{Id}$ in $\Omega \setminus C_\varepsilon$;
**ii)** $\alpha_\delta(x) \in \Delta_0 \quad \forall x \in D_\delta$;
**iii)** for a.e. $x \in \Omega$, $|\nabla\alpha_\delta(x)| \leq (1+3\varepsilon)$;
**iv)** $|\alpha_\delta(x) - x| \leq \varepsilon \ \forall x \in \Omega$.

We want now to define a Borel map $r : \Theta \to \Theta$ such that $r(\theta)$ is a path with the same endpoints as $\theta$ and with the property that, if $\mathscr{H}^1(\theta) \leq L$ and $n$ is large enough,

$$\bar{\delta}_{\Sigma_{n,\varepsilon}}\big(r(\theta)\big) \leq \delta_{\Sigma_n}(\theta) + K(\varepsilon) \tag{4.19}$$

for some $K(\varepsilon) \underset{\varepsilon\to 0}{\longrightarrow} 0$. To do that, we set

$$r(\theta) := \theta_1 \cdot (\alpha_\delta \circ \theta) \cdot \theta_2 \,,$$

where $\theta_1$ and $\theta_2$ are the segments connecting $\theta(0)$ with $\alpha_\delta(\theta(0))$ and $\alpha_\delta(\theta(1))$ with $\theta(1)$ respectively. Therefore, by construction and recalling the properties iii) and iv) above, one has

$$\mathscr{H}^1\big(r(\theta)\big) \leq 2\varepsilon + \mathscr{H}^1(\theta) + 3L\varepsilon = \mathscr{H}^1(\theta) + K_3(\varepsilon)\,, \tag{4.20}$$

where again $K_3(\varepsilon) \underset{\varepsilon\to 0}{\longrightarrow} 0$. Write now $\theta = \theta_a \cup \theta_b \cup \theta_c$, where

$$\theta_a \subseteq \Omega \setminus C_\varepsilon\,, \qquad \theta_b \subseteq C_\varepsilon \setminus D_\delta\,, \qquad \theta_c \subseteq D_\delta \;:$$

we consider now separately $\theta_a$, $\theta_b$ and $\theta_c$. Concerning $\theta_a$, by the property i) we have $\alpha_\delta(\theta_a) = \theta_a$, so that by (4.16)

$$\mathscr{H}^1\big(\alpha_\delta(\theta_a) \setminus \Sigma_{n,\varepsilon}\big) = \mathscr{H}^1\big(\theta_a \setminus \Sigma_n\big)\,. \tag{4.21}$$

Concerning $\theta_b$, by (4.18) we know that

$$\mathscr{H}^1(\Sigma_n \cap \theta_b) = \mu_n(\theta_b) \leq 2K_2(\varepsilon)$$

then, also by property iii), one has

$$\begin{aligned}\mathscr{H}^1\big(\alpha_\delta(\theta_b) \setminus \Sigma_{n,\varepsilon}\big) &\leq (1+3\varepsilon)\mathscr{H}^1(\theta_b) \leq \mathscr{H}^1(\theta_b) + 3L\varepsilon \\ &\leq \mathscr{H}^1\big(\theta_b \setminus \Sigma_n\big) + 2K_2(\varepsilon) + 3L\varepsilon\,.\end{aligned} \tag{4.22}$$

Finally, concerning $\theta_c$, we know by ii) that

$$\alpha_\delta(\theta_c) \subseteq \Delta_0 \subseteq \Sigma_{n,\varepsilon}$$

so that

$$\mathscr{H}^1\big(\alpha_\delta(\theta_c) \setminus \Sigma_{n,\varepsilon}\big) = 0\,. \tag{4.23}$$

Recalling now that

$$r(\theta) = \alpha_\delta(\theta_a) \cup \alpha_\delta(\theta_b) \cup \alpha_\delta(\theta_c) \cup \theta_1 \cup \theta_2$$

and putting together (4.21), (4.22) and (4.23), we derive that

$$\mathscr{H}^1\big(r(\theta) \setminus \Sigma_{n,\varepsilon}\big) \leq \mathscr{H}^1(\theta \setminus \Sigma_n) + K_4(\varepsilon)\,,$$

where again $K_4 \xrightarrow[\varepsilon\to 0]{} 0$. As a consequence,

$$0 \le l := \Big(\mathscr{H}^1(\theta\setminus\Sigma_n) - \mathscr{H}^1\big(r(\theta)\setminus\Sigma_{n,\varepsilon}\big) + K_3(\varepsilon) + K_4(\varepsilon)\Big)\wedge\mathscr{H}^1\big(r(\theta)\cap\Sigma_{n,\varepsilon}\big)$$
$$\le \mathscr{H}^1\big(r(\theta)\cap\Sigma_{n,\varepsilon}\big)\,.$$

Making use of the formula for $\bar\delta_\Sigma$ given in Proposition 2.18 and also by (4.20), we deduce (recall that $\mathscr{H}^1(\theta)\le L$ and denote as usual by $\omega$ the modulus of continuity of $A$ in $[0,L]$)

$$\begin{aligned}\bar\delta_{\Sigma_{n,\varepsilon}}\big(r(\theta)\big) &\le A\big(\mathscr{H}^1\big(r(\theta)\setminus\Sigma_{n,\varepsilon}\big)+l\big) + B\big(\mathscr{H}^1\big(r(\theta)\cap\Sigma_{n,\varepsilon}\big)-l\big)\\ &\le A\big(\mathscr{H}^1(\theta\setminus\Sigma_n)+K_3(\varepsilon)+K_4(\varepsilon)\big)\\ &\qquad + B\Big(0\vee\Big(\mathscr{H}^1(r(\theta)) - \mathscr{H}^1(\theta\setminus\Sigma_n) - K_3(\varepsilon)-K_4(\varepsilon)\Big)\Big)\\ &\le A\big(\mathscr{H}^1(\theta\setminus\Sigma_n)\big)+\omega\big(K_3(\varepsilon)+K_4(\varepsilon)\big)\\ &\qquad + B\Big(0\vee\Big(\mathscr{H}^1(\theta\cap\Sigma_n)-K_4(\varepsilon)\Big)\Big)\\ &\le \delta_{\Sigma_n}(\theta)+\omega\big(K_3(\varepsilon)+K_4(\varepsilon)\big)\ :\end{aligned}$$

then we finally proved (4.19) with

$$K(\varepsilon) := \omega\big(K_3(\varepsilon)+K_4(\varepsilon)\big)\,.$$

Take now any t.p.m. $\eta$ such that $\mathscr{H}^1(\theta)\le L$ for all $\theta\in\operatorname{spt}\eta$. By construction $r_\#\eta$ is another t.p.m., so recalling Proposition 2.14 and thanks to (4.19) we derive

$$MK(\Sigma_{n,\varepsilon}) \le \overline{C}_{\Sigma_{n,\varepsilon}}(r_\#\eta) \le C_{\Sigma_n}(\eta)+K(\varepsilon)\,.$$

As we already noticed in Lemma 4.6, one can take such a t.p.m. $\eta$ with $C_{\Sigma_n}(\eta)$ arbitrarily close to $MK(\Sigma_n)$, therefore we deduce

$$MK(\Sigma_{n,\varepsilon}) \le MK(\Sigma_n)+K(\varepsilon)\,. \tag{4.24}$$

Due to (4.15),

$$\lim\mu_n(C_\varepsilon) = \mu(C_\varepsilon) \ge \mu(\Delta_0) > \mathscr{H}^1(\Delta_0)\,.$$

Hence, at least for $n$ large enough, one has

$$\mathscr{H}^1(\Sigma_{n,\varepsilon}) < \mathscr{H}^1(\Sigma_n)$$

(recall the definition (4.16) of $\Sigma_{n,\varepsilon}$). Together with (4.24), this ensures that for $n$ large enough one has

$$\mathfrak{F}(\Sigma_{n,\varepsilon}) \le \mathfrak{F}(\Sigma_n)+K(\varepsilon)\,.$$

Up to a subsequence, we may assume $\mu_{n,\varepsilon} \xrightarrow[n\to\infty]{*} \mu_\varepsilon$. Thus, minding that

$$\mu_{n,\varepsilon} \le \mu_n \vee (\mathscr{H}^1 \llcorner \Delta_0)\,,$$

we get

$$\mu_\varepsilon \le \mu \vee (\mathscr{H}^1 \llcorner \Delta_0)\,.$$

In particular,

$$\mu_\varepsilon(\Omega \setminus C_\varepsilon) \le \mu(\Omega \setminus C_\varepsilon)\,,$$

and also $\mu_\varepsilon(\partial C_\varepsilon) = 0$. On the other hand, the latter fact implies

$$\mu_\varepsilon(C_\varepsilon) = \lim_{n\to\infty} \mu_{n,\varepsilon}(C_\varepsilon) = \mathscr{H}^1(\Delta_0)\,,$$

while $\mu(C_\varepsilon) \ge \mu(\Delta_0)$. Therefore,

$$\begin{aligned} \|\mu_\varepsilon\| &\le \mu_\varepsilon(\Omega \setminus C_\varepsilon) + \mathscr{H}^1(\Delta_0) \le \|\mu\| - \mu(C_\varepsilon) + \mathscr{H}^1(\Delta_0) \\ &\le \|\mu\| - \mu(\Delta_0) + \mathscr{H}^1(\Delta_0)\,. \end{aligned} \tag{4.25}$$

Finally, by the lower semicontinuity of $\mathfrak{F}$ and since $\mathfrak{F}(\mu_n) \to \mathfrak{F}(\mu)$, we know that

$$\mathfrak{F}(\mu_\varepsilon) \le \liminf_{n\to\infty} \mathfrak{F}(\mu_{n,\varepsilon}) \le \liminf_{n\to\infty} \mathfrak{F}(\mu_n) + K(\varepsilon) = \mathfrak{F}(\mu) + K(\varepsilon)\,.$$

Let us call now $\nu_0$ a weak* limit of some sequence $\mu_{\varepsilon_j}$ with $\varepsilon_j \to 0$ for $j \to \infty$: since $K(\varepsilon_j) \to 0$ we have, again by the lower semicontinuity of $\mathfrak{F}$, that $\mathfrak{F}(\nu_0) \le \mathfrak{F}(\mu)$, and hence $\nu_0$ is optimal. On the other hand, since

$$\mu_\varepsilon \le \mu \vee (\mathscr{H}^1 \llcorner \Delta_0)\,,$$

by (4.25) and the lower semicontinuity of the norm one has

$$\nu_0 \le \mu \vee (\mathscr{H}^1 \llcorner \Delta_0)\,, \qquad \|\nu_0\| \le \|\mu\| - \mu(\Delta_0) + \mathscr{H}^1(\Delta_0)\,. \tag{4.26}$$

Summarizing, starting from a Lipschitz path $\Delta_0$ such that $\mu(\Delta_0) > \mathscr{H}^1(\Delta_0)$, we constructed a measure $\nu_0$ which is optimal for $\mathfrak{F}$ and verifies (4.26). In the very same way, the construction above also works if $\Delta_0$ is replaced by a finite union of disjoint Lipschitz paths. We are then going to define a sequence $\{\Delta_k\}_{k\in\mathbb{N}}$ of finite unions of disjoint Lipschitz paths contained in $\Delta_0$: to this aim, for every integer $k$, we divide $\Delta_0 = \widetilde{PQ}$ in $k$ subpaths $\Delta_i^k$, $i = 1, \ldots, k$ of the same length $\mathscr{H}^1(\widetilde{PQ})/k$, and we define $\Delta_k$ as the union of all $\Delta_i^k$ for which

$$\mu(\Delta_i^k) > \mathscr{H}^1(\Delta_i^k)\,.$$

As already said, we can repeat the construction above with $\Delta_k$ in place of $\Delta_0$, obtaining a measure $\nu_k$ which is optimal for $\mathfrak{F}$ and such that

$$\begin{aligned} &\nu_k \leq \mu \vee (\mathscr{H}^1 \llcorner \Delta_k)\,, \\ &\|\nu_k\| \leq \|\mu\| - \mu(\Delta_k) + \mathscr{H}^1(\Delta_k) \leq \|\mu\| - \mu(\Delta_0) + \mathscr{H}^1(\Delta_0)\,, \end{aligned} \tag{4.27}$$

where the last inequality follows from the fact that

$$\mu(\Delta_0 \setminus \Delta_k) \leq \mathscr{H}^1(\Delta_0 \setminus \Delta_k)\,.$$

Finally, passing to the limit as $k \to \infty$, there is a measure $\nu$ such that, up to a subsequence, $\nu_k \xrightarrow{*} \nu$. By lower semicontinuity the measure $\nu$ is optimal for $\mathfrak{F}$; moreover, writing $\mu = \varphi \mathscr{H}^1 \llcorner \Sigma$ according to Lemma 4.10, by construction one has

$$\mathscr{H}^1 \llcorner \Delta_k \xrightarrow{*} \mathscr{H}^1 \llcorner \{x \in \widetilde{PQ} :\ \varphi(x) > 1\}\,.$$

Hence, the limit in (4.27) gives

$$\nu \leq \mu \vee (\mathscr{H}^1 \llcorner \{\varphi(x) > 1\}) = \mu, \quad \|\nu\| \leq \|\mu\| - \mu(\Delta_0) + \mathscr{H}^1(\Delta_0) < \|\mu\|\,,$$

so we found the desired contradiction to the fact that $\mu$ is a minimal optimal measure. □

We give now a useful representation of $\mathfrak{F}(\mu)$ for a measure $\mu \in \mathcal{M}_1^+(\Omega)$.

**Proposition 4.15.** *For any $\mu \in \mathcal{M}_1^+(\Omega)$ one has*

$$\mathfrak{F}(\mu) = MK(\mu) + H(\|\mu\|)\,, \tag{4.28}$$

*where*

$$MK(\mu) := \inf\left\{C_\mu(\eta) \ :\ \eta \text{ is a t.p.m.}\right\}, \tag{4.29}$$

$$C_\mu(\eta) := \int_\Theta \delta_\mu(\theta)\, d\eta(\theta)\,, \tag{4.30}$$

$$\delta_\mu(\theta) := A\big(\mathscr{H}^1(\theta) - \mu(\theta)\big) + B\big(\mu(\theta)\big)\,. \tag{4.31}$$

*Remark 4.16.* We point out that equations (4.28), (4.29), (4.30) and (4.31) are the generalizations of (2.11), (2.9), (2.8) and (2.2) respectively. Note also that, in order for these equations to make sense, it is necessary that $\mu \in \mathcal{M}_1^+(\Omega)$, i.e. $\mu = \varphi \mathscr{H}^1 \llcorner \Sigma$ with $\varphi \leq 1$. Otherwise, if $\mu$ has parts of dimension lower than one, or with dimension one but density greater than one, then $\mathscr{H}^1(\theta) - \mu(\theta)$ could be negative, which is meaningless. On the other hand, if $\mu$ has parts of dimension greater than one then those parts have no effect in (4.31), and so we implicitly need to make use of Lemma 4.11. We also remark that the above result is extremely useful: in fact, it allows to evaluate the cost $\mathfrak{F}$ of any measure $\mu$ making use of the transport path measures, exactly as one does for the sets $\Sigma$, instead of working with optimal sequences of sets as we needed to do until now.

We may now prove Proposition 4.15.

*Proof (of Proposition 4.15).* Having taken (4.29), (4.30) and (4.31) as definitions, we need to establish (4.28).

First of all, it is useful to compute, exactly as in (2.13) and (2.21), the relaxed envelope $\bar{\delta}_\mu$ of $\delta_\mu$ with fixed endpoints, and the corresponding generalized cost $\overline{C}_\mu$, as well as to generalize the definition (2.5) of the distance in $\Omega$ and of the cost $I_\Sigma$ in the obvious way:

$$\begin{aligned} \bar{\delta}_\mu(\theta) &:= \inf\left\{\liminf_{n\to\infty} \delta_\mu(\theta_n) :\ \theta_n(0)=\theta(0),\ \theta_n(1)=\theta(1),\ \theta_n \xrightarrow{\Theta} \theta\right\}; \\ \overline{C}_\mu(\eta) &:= \int_\Theta \bar{\delta}_\mu(\theta)\, d\eta(\theta); \\ d_\mu(x,y) &:= \inf\left\{\delta_\mu(\theta) :\ \theta\in\Theta,\ \theta(0)=x,\ \theta(1)=y\right\}; \\ I_\mu(\gamma) &:= \iint_{\Omega\times\Omega} d_\mu(x,y)\, d\gamma(x,y). \end{aligned} \tag{4.32}$$

Then, as in Proposition 2.14, one shows that

$$\begin{aligned} MK(\mu) &= \min\left\{I_\mu(\gamma) \ :\ \gamma \text{ is a transport plan}\right\} \\ &= \inf\left\{C_\mu(\eta) :\ \eta \text{ is a t.p.m.}\right\} \\ &= \min\left\{\overline{C}_\mu(\eta) :\ \eta \text{ is a t.p.m.}\right\}; \end{aligned} \tag{4.33}$$

moreover, as in Proposition 2.18 one has

$$\begin{aligned} \bar{\delta}_\mu(\theta) &= \inf\left\{A\big(\mathscr{H}^1(\theta)-\mu(\theta)+l\big)+B\big(\mu(\theta)-l\big) :\ 0\le l\le \mu(\theta)\right\} \\ &= \inf\left\{A\big(\mathscr{H}^1(\theta)-l\big)+B\big(l\big) :\ 0\le l\le \mu(\theta)\right\}. \end{aligned} \tag{4.34}$$

Consider now any sequence of closed sets $\{\Sigma_n\}$ such that $\mathscr{H}^1 \llcorner \Sigma_n \stackrel{*}{\rightharpoonup} \mu$: given any sequence of paths $\theta_n \to \theta$, we claim that

$$\bar{\delta}_\mu(\theta) \le \liminf_{n\to\infty} \bar{\delta}_{\Sigma_n}(\theta_n). \tag{4.35}$$

In fact, by Gołąb Theorem we know that

$$\mathscr{H}^1(\theta) \le \liminf \mathscr{H}^1(\theta_n),$$

and on the other hand by Lemma 4.1 one has

$$\mu(\theta) \ge \limsup \mathscr{H}^1 \llcorner \Sigma_n(\theta_n).$$

This gives (4.35), since recalling (4.34) we get

$$
\begin{aligned}
\bar{\delta}_\mu(\theta) &= \inf\Big\{A\big(\mathscr{H}^1(\theta)-l\big)+B\big(l\big) :\ 0\le l\le \mu(\theta)\Big\}\\
&\le \liminf_{n\to\infty}\ \inf\Big\{A\big(\mathscr{H}^1(\theta)-l\big)+B\big(l\big) :\ 0\le l\le \mu(\theta)\wedge\mathscr{H}^1(\theta_n)\Big\}\\
&\le \liminf_{n\to\infty}\ \inf\Big\{A\big(\mathscr{H}^1(\theta_n)-l\big)+B\big(l\big) :\ 0\le l\le \mu(\theta)\wedge\mathscr{H}^1(\theta_n)\Big\}\\
&\le \liminf_{n\to\infty}\ \inf\Big\{A\big(\mathscr{H}^1(\theta_n)-l\big)+B\big(l\big) :\ 0\le l\le \mu(\theta)\wedge\mathscr{H}^1 \llcorner \Sigma_n(\theta_n)\Big\}\\
&= \liminf_{n\to\infty}\ \inf\Big\{A\big(\mathscr{H}^1(\theta_n)-l\big)+B\big(l\big) :\ 0\le l\le \mathscr{H}^1 \llcorner \Sigma_n(\theta_n)\Big\}\\
&= \liminf_{n\to\infty} \bar{\delta}_{\Sigma_n}(\theta).
\end{aligned}
$$

Hence, we proved the $\Gamma-\liminf$ inequality (4.35). Take now any sequence of t.p.m.'s $\eta_n$ optimal for $\overline{C}_{\Sigma_n}$, and let $\eta$ be a weak* limit (possibly, up to a subsequence) of $\{\eta_n\}$. By Lemma B.20, we have

$$
\begin{aligned}
MK(\mu)\le \overline{C}_\mu(\eta) &= \int_\Theta \bar{\delta}_\mu(\theta)\,d\eta(\theta)\\
&\le \int_\Theta \liminf_{n\to\infty} \bar{\delta}_{\Sigma_n}(\theta)\,d\eta(\theta) \le \liminf_{n\to\infty}\int_\Theta \bar{\delta}_{\Sigma_n}(\theta)\,d\eta_n(\theta)\\
&= \liminf_{n\to\infty} \overline{C}_{\Sigma_n}(\eta_n) = \liminf_{n\to\infty} MK(\Sigma_n).
\end{aligned}
$$

Since it is also true that

$$H(\|\mu\|)\le \liminf H\big(\mathscr{H}^1(\Sigma_n)\big)$$

because $H$ is l.s.c. and nondecreasing and since

$$\|\mu\|\le \liminf \|\mathscr{H}^1 \llcorner \Sigma_n\| = \liminf \mathscr{H}^1(\Sigma_n)\,,$$

it follows that

$$MK(\mu)+H(\|\mu\|)\le \liminf_{n\to\infty} MK(\Sigma_n)+H(\|\Sigma_n\|) = \liminf_{n\to\infty} \mathfrak{F}(\Sigma_n)\,.$$

Recalling the definition (4.1) of $\mathfrak{F}$ for general measures, we get

$$\mathfrak{F}(\mu)\ge MK(\mu)+H(\|\mu\|)\,,$$

so that the first inequality in (4.28) is shown.

We show now the opposite inequality: thanks to Lemma 4.18 below, there exists an optimal t.p.m. $\eta$ for $\mu$ (that is, a t.p.m. $\eta$ with $\overline{C}_\mu(\eta)=MK(\mu)$) with $\operatorname{spt}\eta$ contained in the subset of $\Theta$ made by all the paths of length bounded by $L$. Then, since $\mu\in\mathcal{M}_1^+(\Omega)$, we can write $\mu=\varphi\mathscr{H}^1 \llcorner \Sigma$, and we can consider $\Sigma$ to be the countable union $\cup_i\theta_i$ where each $\theta_i$ is a Lipschitz path of length bounded by $L$. We define then $\Sigma_n$ as follows: we take the paths $\theta_i$ for $i=1, 2, \ldots, n$, we divide each of them in the union of $n$ subpaths

$\theta_{i,1}, \theta_{i,2}, \ldots, \theta_{i,n}$ of length $\mathscr{H}^1(\theta_i)/n$, and finally we let $\Sigma_n$ be the union of $n^2$ closed connected subpaths, each of which is contained in $\theta_{i,j}$ for $(i,j) \in \{1, 2, \ldots, n\}^2$ and with length $\mu(\theta_{i,j})$. Note that this is possible thanks to the hypothesis $\varphi \leq 1$, since it implies $\mathscr{H}^1(\theta_{i,j}) \geq \mu(\theta_{i,j})$.

One notices that the characteristic functions $\chi_{\Sigma_n} : \Sigma \to \mathbb{R}$ converge to $\varphi$ weakly$*$ in $L^\infty(\Sigma)$, thus for any $\theta \in \Theta$ one has

$$\mathscr{H}^1 \llcorner \Sigma_n(\theta) = \int_\Sigma \chi_{\Sigma_n}(s)\chi_\theta(s)\, d\mathscr{H}^1(s) \to \int_\Sigma \varphi(s)\chi_\theta(s)\, d\mathscr{H}^1(s) = \mu(\theta)\,,$$

and hence

$$\limsup_{n\to\infty} \bar{\delta}_{\Sigma_n}(\theta) \leq \bar{\delta}_\mu(\theta)\,.$$

We use now the Dominated Convergence Theorem, which is possible since, on the set of paths of length less than $L$, $\bar{\delta}_{\Sigma_n}$ is bounded by a constant that does not depend on $n$ (indeed, $\bar{\delta}_{\Sigma_n}(\theta) \leq A\big(\mathscr{H}^1(\theta)\big) \leq A(L)$); we obtain that

$$\begin{aligned}\liminf_{n\to\infty} MK(\Sigma_n) &\leq \liminf_{n\to\infty} \overline{C}_{\Sigma_n}(\eta) \\ &= \liminf_{n\to\infty} \int_\Theta \bar{\delta}_{\Sigma_n}(\theta)\, d\eta \leq \int_\Theta \limsup_{n\to\infty} \bar{\delta}_{\Sigma_n}(\theta)\, d\eta \\ &\leq \int_\Theta \bar{\delta}_\mu(\theta)\, d\eta = \overline{C}_\mu(\eta) = MK(\mu)\,.\end{aligned} \tag{4.36}$$

Since by construction

$$\mathscr{H}^1(\Sigma_n) = \|\mu\| \qquad \forall n \in \mathbb{N}$$

and $\mathscr{H}^1 \llcorner \Sigma_n \stackrel{*}{\rightharpoonup} \mu$, from (4.36) it follows

$$\begin{aligned}\mathfrak{F}(\mu) &\leq \liminf_{n\to\infty} \mathfrak{F}(\Sigma_n) = \liminf_{n\to\infty} MK(\Sigma_n) + H\big(\mathscr{H}^1(\Sigma_n)\big) \\ &= \liminf_{n\to\infty} MK(\Sigma_n) + H(\|\mu\|) \\ &\leq MK(\mu) + H(\|\mu\|)\,.\end{aligned}$$

The second inequality in (4.28) is then shown, so the proof is complete. □

We prove now the generalizations to the relaxed functional $\mathfrak{F}$ of some properties that we already encountered for the original functional in Proposition 2.3, Lemma 2.16, Corollary 2.17 and Proposition 4.3. All these properties can be obtained in a very similar way to those of the original functional. It is convenient to introduce for this purpose the new set of measures

$$\mathcal{M}^+_{\mu,0}(\Theta) := \operatorname{Arg\,min}\{\overline{C}_\mu(\eta) \;:\; \eta \text{ a t.p.m.}\}\,.$$

**Lemma 4.17.** *For any measure $\mu \in \mathcal{M}^+_1(\Omega)$, the following properties hold:*
*i) the distance $d_\mu$ is continuous;*

*ii) there exists an* optimal t.p.m., *that is a t.p.m.* $\eta_{\rm opt} \in \mathcal{M}^+_{\mu,0}(\Theta)$*;*
*iii) for any* $\eta_{\rm opt} \in \mathcal{M}^+_{\mu,0}(\Theta)$*, one has that* $\operatorname{spt} \eta_{\rm opt}$ *is contained in the set of all geodesics with respect to* $d_\mu$*.*

*Proof.* By construction, the function $\theta \mapsto \bar{\delta}_\mu(\theta)$ is lower semicontinuous: as a consequence, for any pair $(x, y) \in \Omega \times \Omega$ the distance $d_\mu(x, y)$, introduced in (4.32), can be equivalently defined as the minimum of $\bar{\delta}_\mu(\theta)$ among the paths $\theta \in \Theta$ connecting $x$ and $y$; as in the classical case, we will call *geodesics* these minimizers. Recalling again the lower semicontinuity of $\bar{\delta}_\mu$, it follows also immediately that the set $G$ of the geodesics is a closed subset of $\Theta$, as well as the lower semicontinuity of $d_\mu$. The proof of the upper semicontinuity of $d_\mu$ is identical to the classical case, making use of the continuity of $A$ and of Lemma 2.1 (which clearly holds also when $\Sigma$ is a countable union of paths of uniformly bounded length); hence, i) holds.

Concerning the existence of an optimal t.p.m. , this is also obtained exactly as in the classical case: the existence of an optimal transport plan $\gamma_{\rm opt}$, i.e. a measure $\gamma$ minimizing $I_\mu(\gamma)$ as defined in (4.32), is again standard. Moreover, the closedness of $G$ and the existence of geodesics for each pair in $\Omega \times \Omega$ ensures again the existence of a Borel map

$$q : \Omega \times \Omega \to \Theta$$

associating to any pair of points a geodesics connecting them; finally, the t.p.m. $q_{\#}\gamma_{\rm opt}$ is clearly an optimal t.p.m. thanks to (4.33); therefore, also ii) is shown.

Finally, the property iii) is proved again as in the classical case: thanks to (4.33), any optimal t.p.m. is concentrated in the set $G$ of the geodesics; therefore, since this set is closed as already remarked, the whole support of any optimal t.p.m. is contained in $G$. □

**Lemma 4.18.** *Given any* $\mu \in \mathcal{M}^+_1(\Omega)$ *there exists a constant* $L$ *depending only on* $A$*,* $\Omega$ *and* $\|\mu\|$ *such that for any pair* $(x, y) \in \Omega \times \Omega$ *the Euclidean length* $\mathscr{H}^1(\theta)$ *of some geodesic* $\theta$ *connecting* $x$ *and* $y$ *(in particular, of* each *geodesic if* $A$ *is not constant for all values large enough) is bounded by* $L$*. In addition, there exists an optimal t.p.m.* $\eta$ *for* $\overline{C}_\mu$ *(in particular,* each *optimal t.p.m. if* $A$ *is not constant for all values large enough) such that* $\operatorname{spt} \eta$ *is contained in the set of paths* $\theta$ *with Euclidean length bounded by* $L$*.*

*Proof.* For the first part, one can simply adapt the proof of Lemma 2.16 to this case; one only has to notice, as we already did, that Lemma 2.1 is true also in the case when $\Sigma$ is the union of countably many paths of equibounded length.

The proof of the second part, also in light of Lemma 4.17-iii), can be made exactly as the proof of Corollary 2.17. □

**Proposition 4.19.** *Given* $\mu \in \mathcal{M}^+_1(\Omega)$ *and a sequence* $\{\mu_n\} \in \mathcal{M}^+_1(\Omega)$ *such that* $\mu_n \stackrel{*}{\rightharpoonup} \mu$*, one has that*

$$\bar{\delta}_\mu \leq \Gamma - \liminf \bar{\delta}_{\mu_n} ,$$

*that is*

$$\bar{\delta}_\mu(\theta) \leq \liminf \bar{\delta}_{\mu_n}(\theta_n)$$

*whenever* $\theta_n \to \theta$ *uniformly. As a consequence,* $d_\mu \leq \liminf d_{\mu_n}$.

*Proof.* Take any sequence $\theta_n$ of paths uniformly converging to some path $\theta$; therefore,

$$\mu(\theta) \geq \limsup \mu_n(\theta_n)$$

by Lemma 4.1, and on the other hand

$$\mathscr{H}^1(\theta) \leq \liminf \mathscr{H}^1(\theta_n)$$

by Gołąb Theorem. We can argue exactly as when we showed property (4.35) in the proof of Proposition 4.15 to obtain

$$\bar{\delta}_\mu(\theta) \leq \liminf \bar{\delta}_{\mu_n}(\theta_n) ,$$

so the first part of the proof follows.

Concerning the second one, given any pair $(x, y) \in \Omega \times \Omega$, we can choose according to Lemma 4.18 paths $\theta_n$ having Euclidean length less than $L$ and being geodesics with respect to $d_{\mu_n}$, that is $\bar{\delta}_{\mu_n}(\theta_n) = d_{\mu_n}(x, y)$. As usual, we can assume up to a subsequence that $\theta_n \to \theta$ uniformly as elements of $C([0, 1], \Omega)$ and in the Hausdorff distance as closed subsets of $\Omega$. Recalling the $\Gamma$−liminf inequality shown above we conclude, since

$$d_\mu(x, y) \leq \bar{\delta}_\mu(\theta) \leq \liminf_{n\to\infty} \bar{\delta}_{\mu_n}(\theta_n) = \liminf_{n\to\infty} d_{\mu_n}(x, y) .$$

□

We remark now what follows: in Lemmas 4.9 and 4.11 and Theorem 4.14, we showed the existence of some optimal measure respectively which is absolutely continuous with respect to $\mathscr{H}^1$, which is concentrated in some countable union of curves of uniformly bounded length, and which belongs to $\mathcal{M}_1^+(\Omega)$. In fact, we were able to prove that the above properties are satisfied by any minimal optimal measure. Roughly speaking, if $\mu$ is an optimal measure and $\mu' \leq \mu$ is minimal and optimal, we proved that the part $\mu - \mu'$ of the optimal measure is "useless"; formally, we can say that the part $\mu - \mu' \leq \mu$ is *useless* if $d_\mu \equiv d_{\mu'}$, and then $MK(\mu) = MK(\mu')$ and $\mathfrak{F}(\mu) \geq \mathfrak{F}(\mu')$. It is now quite easy to guess that, often, a stronger assertion than the one of the above lemmas is true, namely that *all* the optimal measures possess the desired properties: this holds whenever the presence of a useless part of an optimal measure is impossible, that is, when each optimal measure is minimal. This is certainly true, for instance, in the two situations considered in the next proposition.

**Proposition 4.20.** *All the optimal measures are minimal in each of the following cases:*

- *the function $H$ is* strictly *increasing;* (4.37)
- *$B'(s) < A'(t)$ for every $s \in \mathbb{R}^+$, $0 \le t \le \mathrm{diam}(\Omega)$ and $f^+ - f^-$ is not concentrated on a set of finite $\mathscr{H}^1$-measure.* (4.38)

*Proof.* The first situation is clear: indeed, as noticed before, if there is a non-minimal optimal measure $\mu$, and so a non-null useless part $\mu - \mu'$, then $MK(\mu) = MK(\mu')$; by the strict monotonicity of $H$, $H(\mu') < H(\mu)$, which leads to the contradiction $\mathfrak{F}(\mu') < \mathfrak{F}(\mu)$.

Let us then consider the case (4.38), and assume by contradiction the existence of a non-minimal optimal measure $\mu$, and then the existence of an optimal measure $\mu'$ strictly less than $\mu$: take then an optimal t.p.m. $\eta$ and notice that the conclusion is proved if the set of the paths $\theta$ such that $\mu'(\theta) < \mathscr{H}^1(\theta)$ is not $\eta$–negligible. Indeed, in this case, we could replace $\mu - \mu'$ by the measure $\mathscr{H}^1 \llcorner \theta_0 - \mu \llcorner \theta_0$, for a suitable $\theta_0 \in \Theta$ satisfying

$$0 < \mathscr{H}^1(\theta_0) - \mu(\theta_0) \le \|\mu - \mu'\| .$$

Hence, thanks to the assumption $B'(s) < A'(t)$, this would provide a strictly positive gain because an $\eta$-nonnegligible set of paths can be changed so as to strictly decrease their costs (this can be made rigorous arguing as in Lemma 6.9 below). We can then assume that for $\eta$–a.a. paths $\theta$ we have $\mu'(\theta) = \mathscr{H}^1(\theta)$ so that, denoting by $\Sigma$ the set where $\mu'$ has $\mathscr{H}^1$–density equal to 1, one has $\mathscr{H}^1(\theta) = \mathscr{H}^1(\theta \cap \Sigma)$; the contradiction with (4.38) will now follow by proving that $f^+ - f^-$ is concentrated on a set of finite length. To this aim, we select paths $\theta_n \in \Theta$ almost maximizing

$$\mathscr{H}^1\left(\theta \setminus \cup_{i=0}^{n-1}\theta_i\right)$$

among all paths for which $\mathscr{H}^1(\theta \setminus \Sigma) = 0$, i.e.

$$\begin{aligned} &\mathscr{H}^1(\theta_n \setminus \Sigma) = 0 ,\\ &\mathscr{H}^1\left(\theta_n \setminus \cup_{i=0}^{n-1}\theta_i\right) \ge \sup\left\{\mathscr{H}^1\left(\theta \setminus \cup_{i=0}^{n-1}\theta_i\right),\ \mathscr{H}^1(\theta \setminus \Sigma) = 0\right\} - \frac{1}{2^n} . \end{aligned}$$

Setting now $\Sigma' := \cup\theta_n$, up to adding to $\Sigma$ the $\mathscr{H}^1$–negligible set $\cup_n\theta_n \setminus \Sigma$, we can assume that $\Sigma' \subseteq \Sigma$. Moreover, by construction of $\Sigma'$, we have that

$$\begin{aligned} &\mathscr{H}^1(\Sigma') \le \mathscr{H}^1(\Sigma) \le \|\mu'\| < +\infty ,\\ &\mathscr{H}^1(\theta \setminus \Sigma') = 0 \text{ for } \eta\text{–a.a. } \theta \in \Theta . \end{aligned} \tag{4.39}$$

Notice now that $\Sigma'$ is a countable union of pairwise disjoint and pathwise connected components $\Sigma_n$ (each $\Sigma_n$ is the union of some of the paths $\theta_n$).

Now, we say that the pair $(\Sigma_i, \Sigma_j)$ is *connectible* if there is a path $\theta_{ij} \in \Theta$ such that

$$\mathscr{H}^1(\theta_{ij} \setminus \Sigma') = 0\,, \qquad \mathscr{H}^1(\theta_{ij} \cap \Sigma_i) > 0\,, \qquad \mathscr{H}^1(\theta_{ij} \cap \Sigma_j) > 0\,.$$

If $(\Sigma_i, \Sigma_j)$ is connectible, adding to $\Sigma'$ the $\mathscr{H}^1-$negligible set $\theta_{ij} \setminus \Sigma'$ we have that $\Sigma_i$ and $\Sigma_j$ belong now to the same connected component of $\Sigma'$. Iterating countably many times this procedure, we end up with a set $\Sigma''$ for which (4.39) still holds; but by construction $\Sigma''$ is made by countably many connected components (since so is $\Sigma'$), still denoted by $\Sigma_i$, and no pair $(\Sigma_i, \Sigma_j)$ is connectible; again up to adding to $\Sigma''$ a $\mathscr{H}^1-$negligible set, we can also assume that all $\Sigma_i$ are closed. As a consequence of (4.39) and of the fact that the different connected components of $\Sigma''$ are not connectible, $\eta-$a.e. path $\theta$ is entirely contained in a single $\Sigma_i$; hence, recalling the definition of t.p.m., we deduce that $f^+ - f^-$ is concentrated on the set

$$\Big\{\theta(0) \cup \theta(1) :\ \exists i \in \mathbb{N},\ \theta \subseteq \Sigma_i\Big\} \subseteq \Sigma''\,,$$

of finite length. The desired contradiction to (4.38) then is found, and the proof is completed. □

Summarizing, we know now the existence of optimal measures $\mu \in \mathcal{M}_1^+(\Omega)$: any such measure can be written as $\mu = \varphi\mathscr{H}^1 \llcorner \Sigma$ with $\Sigma$ a rectifiable Borel set and $\varphi : \Sigma \to [0,1]$ a Borel function. In particular, the measure $\mu$ corresponds to a set if and only if

$$\varphi(x) = 1 \qquad\qquad \text{for } \mu\text{–a.e. } x \in \Omega.$$

**Definition 4.21.** The set

$$\mathcal{M}_2^+(\Omega) \subseteq \mathcal{M}_1^+(\Omega) \subseteq \mathcal{M}^+(\Omega)$$

is defined as the set of all measures $\mathscr{H}^1 \llcorner \Sigma$ for some set $\Sigma$ contained in a countable union of Lipschitz paths of uniformly bounded length. Therefore, $\mu = \varphi\mathscr{H}^1 \llcorner \Sigma$ belongs to $\mathcal{M}_2^+(\Omega)$, if and only if $\varphi(x) = 1$ for $\mu-$a.e. $x \in \Omega$.

As we discussed before Definition 4.12, one can expect the existence of optimal measures contained in $\mathcal{M}_1^+(\Omega) \setminus \mathcal{M}_2^+(\Omega)$. Indeed, as we anticipated, there are situations in which some optimal measure (or even all the optimal measures) are contained in $\mathcal{M}_1^+(\Omega) \setminus \mathcal{M}_2^+(\Omega)$: we give such an example in the next section. In that example, as in many others through the monograph, the situation is basically one-dimensional; this means that $f^\pm$, $\Sigma$ and almost all the paths $\theta$ are contained in the same segment. However, recall that the ambient space is always at least two-dimensional, as we explained in the beginning of the monograph. We point the reader's attention to the importance of this fact: indeed, even in these essentially "one-dimensional" situations, the fact that the ambient space has a higher dimension is crucial; otherwise, the setting of the problem should have been different in order to be meaningful. Roughly speaking, if the space is one-dimensional then one cannot walk "close to the railway" because there is no room to do that; thus, however expensive the train ticket is, the passenger has no choice and is forced to buy it.

## 4.3 Non-existence of Classical Solutions

In this section we present a situation in which no optimal measure $\mu$ corresponds to a set, that is $\mu \notin \mathcal{M}_2^+(\Omega)$ for any optimal measure $\mu$. More precisely, it will appear that whenever $A$ and $B$ are strictly convex, one should expect that no optimal measure corresponds to a set, so that the cases when there is an optimal measure $\mu \in \mathcal{M}_2^+(\Omega)$ will be exceptional.

To do that, we take

$$\Omega = [0,2]^2 \subseteq \mathbb{R}^2 ,$$

we assume that $A = B$ is a strictly convex and $\mathrm{C}^1$ function, that $H(s) = 0$ for all $s \in [0,1]$ and $H(s) > 0$ for any $s > 1$, that

$$f^\pm \ll \mathscr{H}^1 \llcorner \big([0,2] \times \{0\}\big) ,$$

and that

$$f^+\big([0,s] \times \{0\}\big) > f^-\big([0,s] \times \{0\}\big) \qquad \forall\, 0 < s < 2 . \tag{4.40}$$

We will denote for brevity by $f^+$ and $f^-$ also the densities of $f^+$ and $f^-$ with respect to $\mathscr{H}^1$, and we will write for convenience $\bar{z} \equiv (z,0) \in \Omega$ for any real number $z \in [0,2]$; finally, we assume that $f^+$ and $f^-$ are strictly positive inside $[0,2] \times \{0\}$. It is clear that all optimal measures $\mu$ are concentrated in $[0,2] \times \{0\}$: otherwise, projecting $\mu$ on $[0,2] \times \{0\}$ would provide a lower value in the cost $\mathfrak{F}$.

Take now an optimal measure

$$\mu = \varphi(x) \mathscr{H}^1 \llcorner [0,2] \in \mathcal{M}_1^+(\Omega)$$

and notice that, for any $x \leq y$ in $[0,2]$ one has

$$\begin{aligned} d_\mu(\bar{x},\bar{y}) &= \inf_{0 \leq t \leq \mu([\bar{x},\bar{y}])} A\big(y-x-t\big) + B\big(t\big) \\ &= A\big(y-x-l\big) + A\big(l\big), \qquad \text{where} \qquad l = \frac{y-x}{2} \wedge \mu([\bar{x},\bar{y}]) . \end{aligned} \tag{4.41}$$

Since the map

$$t \mapsto A\big(y-x-t\big) + B\big(t\big)$$

has its absolute minimum at $t = (y-x)/2$, from (4.41) we get

$$d_\mu(\bar{x},\bar{y}) \geq A\Big(\frac{y-x}{2}\Big) + B\Big(\frac{y-x}{2}\Big) = 2\,A\Big(\frac{y-x}{2}\Big). \qquad (4.42)$$

Note that in the above calculations we used the convexity of the function $A$; moreover, the inequality (4.42) holds strictly if and only if

$$\mu([\bar{x},\bar{y}]) < \frac{y-x}{2}\,,$$

due to the *strict* convexity of $A$.

We present now a well-known property of the transport problems, namely the *cyclical monotonicity*, which is introduced and shown, for instance, in [1, 36, 60]: this property says that, whenever $\gamma$ is an optimal transport plan,

$$d_\mu(p,q) + d_\mu(p',q') \leq d_\mu(p,q') + d_\mu(p',q) \quad \forall (p,q),\, (p',q') \in \operatorname{spt}\gamma\,. \qquad (4.43)$$

The following property is a consequence of the cyclical monotonicity.

**Lemma 4.22.** *In the hypotheses of this section, for any optimal transport plan $\gamma$ one has*

$$x_1 < x_2 \Longrightarrow y_1 \leq y_2 \qquad \text{whenever} \qquad (\bar{x}_1,\bar{y}_1),\, (\bar{x}_2,\bar{y}_2) \in \operatorname{spt}\gamma\,. \qquad (4.44)$$

*Proof.* First of all, we underline that since $\gamma$ is an optimal transport plan then for any $(x,y) \in \operatorname{spt}\gamma$ one has $x \leq y$ thanks to (4.40): this is a standard and well-known fact, since otherwise one can easily find a contradiction with (4.44). Therefore, to show the claim of the lemma it is sufficient to take

$$0 \leq x_1 < x_2 < y \leq 2$$

and to show that

$$\frac{\partial d_\mu}{\partial y}(\overline{x_1},\overline{y'}) > \frac{\partial d_\mu}{\partial y}(\overline{x_2},\overline{y'})\,, \qquad (4.45)$$

denoting by $\partial d_\mu/\partial y$ the partial derivative of $d_\mu(\bar{x},\bar{y})$ relative to the vector $(1,0)$ in the variable $\bar{y}$. To show (4.45), keeping in mind (4.41) we define

$$l_1 := \frac{y'-x_1}{2} \wedge \mu([\overline{x_1},\overline{y'}])\,, \qquad \tilde{l}_1 := y' - x_1 - l_1\,,$$
$$l_2 := \frac{y'-x_2}{2} \wedge \mu([\overline{x_2},\overline{y'}])\,, \qquad \tilde{l}_2 := y' - x_2 - l_2\,,$$

so that

$$d_\mu(\overline{x_1},\overline{y'}) = A\big(\tilde{l}_1\big) + B\big(l_1\big) \quad \text{and} \quad d_\mu(\overline{x_2},\overline{y'}) = A\big(\tilde{l}_2\big) + B\big(l_2\big)\,. \qquad (4.46)$$

Again using (4.41), recalling that $x_1 < x_2$ it is immediate to deduce that

$$\tilde{l}_1 \geq l_1 \qquad \text{and} \qquad \tilde{l}_2 \geq l_2 \, ; \tag{4.47}$$

$$l_1 \geq l_2 \, , \qquad \tilde{l}_1 \geq \tilde{l}_2 \qquad \text{and at least one of the inequalities is strict:} \tag{4.48}$$

the last assertion is obvious since

$$l_1 + \tilde{l}_1 = y' - x_1 > y' - x_2 = l_2 + \tilde{l}_2 \, .$$

Up to an error $o(\varepsilon)$ we can now write, using (4.41) and (4.48), the following estimates:

$$\begin{aligned} d_\mu(\overline{x_1}, \overline{y' + \varepsilon}) - d_\mu(\overline{x_1}, \overline{y'}) &= \varepsilon\big((1 - c_1)\, A'(\tilde{l}_1) + c_1\, A'(l_1)\big) + o(\varepsilon)\, , \\ d_\mu(\overline{x_2}, \overline{y' + \varepsilon}) - d_\mu(\overline{x_2}, \overline{y'}) &= \varepsilon\big((1 - c_2)\, A'(\tilde{l}_2) + c_2\, A'(l_2)\big) + o(\varepsilon)\, , \end{aligned} \tag{4.49}$$

where the constants $c_1$ and $c_2$ are defined by

$$c_1 := \begin{cases} \varphi(x) & \text{if } l_1 < \tilde{l}_1 \, , \\ \varphi(x) \wedge 1/2 & \text{if } l_1 = \tilde{l}_1 \, , \end{cases} \qquad c_2 := \begin{cases} \varphi(x) & \text{if } l_2 < \tilde{l}_2 \, , \\ \varphi(x) \wedge 1/2 & \text{if } l_2 = \tilde{l}_2 \end{cases}$$

(notice that $0 \leq c_1, c_2 \leq 1$ since $\mu \in \mathcal{M}_1^+(\Omega)$). Finally, we conclude considering separately the three possibilities:

<u>*Case I.*</u> $c_1 = c_2$ .
In this case, (4.45) follows directly from (4.49) recalling (4.46), (4.48) and the fact that $A'$ is strictly increasing since $A$ is strictly convex and $\mathrm{C}^1$.

<u>*Case II.*</u> $c_1 < c_2$ .
As in the first case we have

$$(1 - c_2)\, A'(\tilde{l}_1) + c_2\, A'(l_1) > (1 - c_2)\, A'(\tilde{l}_2) + c_2\, A'(l_2) \, ;$$

with the hypothesis $c_1 < c_2$, by (4.47) we deduce

$$(1 - c_1)\, A'(\tilde{l}_1) + c_1\, A'(l_1) \geq (1 - c_2)\, A'(\tilde{l}_1) + c_2\, A'(l_1) \, ,$$

and therefore (4.45) is proved also in this case.

<u>*Case III.*</u> $c_1 > c_2$ .
Recalling the definition of $c_1$ and $c_2$, this case is possible only if $l_2 = \tilde{l}_2$; but then by (4.47) and (4.48) we infer that

$$\tilde{l}_1 \geq l_1 \geq \tilde{l}_2 = l_2 \, ,$$

so that (4.45) follows immediately by (4.49) and the fact that $A'$ is strictly increasing. □

We point out now another well-known fact, i.e. that there is a unique optimal transport plan $\bar{\gamma}$ between $f^+$ and $f^-$ fulfilling property (4.44), hence by Lemma 4.22 a unique optimal transport plan. Otherwise, there would be two optimal transport plans $\gamma_1$ and $\gamma_2$, a number $x \in [0, 2]$ and two numbers $y_1 < y_2 \in [0, 2]$ such that

$$(\bar{x}, \bar{y}_1) \in \operatorname{spt} \gamma_1, \qquad (\bar{x}, \bar{y}_2) \in \operatorname{spt} \gamma_2 \setminus \operatorname{spt} \gamma_1 .$$

By (4.44) and recalling that $f^+$ and $f^-$ are assumed strictly positive in $[0,2] \times \{0\}$, it follows the existence of some $x' > x$ and $y_1 < y' < y_2$ such that $(\bar{x}', \bar{y}') \in \operatorname{spt} \gamma_1$. By the linearity of the cost (2.6) with respect to $\gamma$, it follows that also

$$\gamma = \frac{\gamma_1 + \gamma_2}{2}$$

is an optimal transport plan; but $(\bar{x}, \bar{y}_2)$ and $(\bar{x}', \bar{y}')$ are two pairs in $\operatorname{spt} \gamma$ against (4.44); therefore, the uniqueness of the optimal transport plan $\bar{\gamma}$ is established.

Set now $t(x) = y$ for $x \in [0,2]$, where $y \in [0,2]$ is the unique number such that

$$f^+([0,x] \times \{0\}) = f^-([0,y] \times \{0\}) .$$

This is a well-defined continuous function thanks to the assumption that $f^-$ is strictly positive, and moreover one has $t(0) = 0$, $t(2) = 2$ and $t(x) > x$ for $0 < x < 2$ by (4.40). The unique optimal transport plan $\bar{\gamma}$ is given by

$$\bar{\gamma} := (\mathrm{Id}, \bar{t})_{\#} f^+ ,$$

where $\bar{t} : [0,2] \times \{0\} \to \Omega$ denotes the map $\bar{t}(\bar{x}) := \overline{t(x)}$. Moreover, from the equality

$$\int_0^x f^+(s)\, ds = \int_0^{t(x)} f^-(s)\, ds$$

and the strict positivity of $f^-$, one has that the derivative $t'(x)$ exists and is strictly positive for a.e. $x \in [0,2]$. In words, the transport plan moves the mass from each point

$$\bar{x} \in [0,2] \times \{0\}$$

to the corresponding point

$$\bar{t}(\bar{x}) = \overline{t(x)} \in [0,2] \times \{0\} \ :$$

the fact that, in this case, the unique optimal transport plan is $\bar{\gamma} = (\mathrm{Id}, \bar{t})_{\#} f^+$, is again well-known (see for instance [1, Theorem 3.1]). Note that we found the unique optimal transport plan without knowing anything about the measure $\mu$: in other words, for any measure $\mu \in \mathcal{M}_1^+(\Omega)$, the same plan $\bar{\gamma}$ is the unique minimizer of

$$I_\mu(\gamma) = \iint_{\Omega \times \Omega} d_\mu(p,q)\, d\gamma(p,q) .$$

Thanks to (4.42) and the fact that $H(1) = 0$, the measure

$$\bar{\mu} := \frac{1}{2} \mathscr{H}^1 \llcorner \left( [0,2] \times \{0\} \right)$$

is optimal, since

$$d_{\bar{\mu}}(\bar{x}, \bar{y}) = 2\,A\Big(\frac{y-x}{2}\Big)$$

for any $0 \le x \le y \le 2$. Then, we have found an optimal measure which is not classical, that is, an optimal measure

$$\bar{\mu} \in \mathcal{M}_1^+(\Omega) \setminus \mathcal{M}_2^+(\Omega)\,;$$

we want now to show more, that is, that there is no classical optimal measure $\mu \in \mathcal{M}_2^+(\Omega)$. First of all, we need to establish the next result.

**Lemma 4.23.** *In the assumptions of this chapter, if $\mu$ is an optimal measure then*

$$\mu\big([\bar{x}, \overline{t(x)}]\big) = \frac{t(x) - x}{2} \qquad \forall\, x \in [0,2]\,. \tag{4.50}$$

*Proof.* Recalling (4.42) and the fact that $\bar{\mu}$ is optimal, since $\mu$ is optimal we immediately know that the inequality

$$\mu\big([\bar{x}, \overline{t(x)}]\big) \ge \frac{t(x) - x}{2} \tag{4.51}$$

holds for any $0 \le x \le 2$; we suppose then the existence of a point $0 < x_0 < 2$ such that the strict inequality

$$\mu\big([\bar{x}_0, \overline{t(x_0)}]\big) - \frac{t(x_0) - x_0}{2} =: \rho > 0$$

holds, and we aim to find a contradiction. Let us define, starting from $x_0$, a two-sided sequence $\{x_z\}$ with $z \in \mathbb{Z}$ by induction setting $x_{h+1} = t(x_h)$ (this defines $x_z$ for any $z \in \mathbb{Z}$ since $t$ is continuous and strictly increasing thus invertible). Since $t$ is strictly increasing in $(0,2)$, we have

$$\lim_{z\to\infty} x_z = 2, \qquad\qquad \lim_{z\to-\infty} x_z = 0\,.$$

By the fact that for any $z \in \mathbb{Z}$ the inequality (4.51) holds, for any $z < 0 < m$ we have

$$\mu\big([\bar{x}_z, \bar{x}_m]\big) \ge \frac{x_m - x_z}{2} + \rho\,.$$

Letting $m \to \infty$ and $z \to -\infty$, the preceding inequality ensures

$$\|\mu\| \ge 1 + \rho\,;$$

but then

$$H(\|\mu\|) > 0 = H(\|\bar{\mu}\|)\,,$$

hence $\mathfrak{F}(\bar{\mu}) < \mathfrak{F}(\mu)$ against the optimality of $\mu$, and we found the desired contradiction. □

We can now prove, as a consequence of (4.50), the claim that we already stated before, that is, that no optimal measure is classical (i.e. belongs to $\mathcal{M}_2^+(\Omega)$).

**Lemma 4.24.** *In the assumptions of this chapter, there are no optimal measures $\mu \in \mathcal{M}_2^+(\Omega)$.*

*Proof.* Take any optimal measure $\mu$, and let $x \in (0,2)$ be a point such that $\bar{x}$ is a point of density 1 of $\mu$ and the derivative $t'(x)$ exists. By (4.50), one has

$$\mu([\bar{x}, \overline{t(x)}]) = \frac{t(x) - x}{2}, \qquad \mu([\overline{x+\varepsilon}, \overline{t(x+\varepsilon)}]) = \frac{t(x+\varepsilon) - (x+\varepsilon)}{2}.$$

This implies

$$\mu\big([\overline{t(x)}, \overline{t(x+\varepsilon)}]\big) - \mu\big([\bar{x}, \overline{x+\varepsilon}]\big) = \frac{t(x+\varepsilon) - t(x) - \varepsilon}{2}.$$

But

$$\mu\big([\bar{x}, \overline{x+\varepsilon}\,]\big) = \varepsilon + o(\varepsilon),$$

because $\bar{x}$ is a point of density 1; moreover

$$t(x+\varepsilon) = t(x) + \varepsilon t'(x) + o(\varepsilon)$$

since $t'(x)$ is defined. It follows

$$\mu\big([\overline{t(x)}, \overline{t(x) + \varepsilon t'(x)}]\big) = \frac{t'(x)+1}{2}\,\varepsilon + o(\varepsilon),$$

and so the density of $\mu$ at $\overline{t(x)}$ is

$$\frac{t'(x)+1}{2\,t'(x)}.$$

Analogously, if the density of $\mu$ at $\bar{x}$ is 0, then the density at $\overline{t(x)}$ is given by

$$\frac{t'(x)-1}{2\,t'(x)}.$$

If the measure $\mu$ belongs to $\mathcal{M}_2^+(\Omega)$, then the density of $\mu$ at a.e. point $\bar{x} \in [0,2] \times \{0\}$ is either 0 or 1; since, as already pointed out, the derivative $t'(x)$ exists and is strictly positive at almost each $x \in [0,2]$, we deduce that it should be $t'(x) = 1$ for a.e. $x \in [0,2]$. But this would imply $t \equiv \mathrm{Id}$, hence $f^+ = f^-$, which gives a contradiction with (4.40), which concludes the proof. □

Summarizing, for any choice of $f^+$ and $f^-$ as in the assumptions of this chapter all the optimal measures do not belong to $\mathcal{M}_2^+(\Omega)$, while for instance the measure

$$\bar{\mu} = \frac{\mathscr{H}^1 \llcorner ([0,2] \times \{0\})}{2} \in \mathcal{M}_1^+(\Omega) \setminus \mathcal{M}_2^+(\Omega)$$

is always optimal.

## 4.4 Existence of Classical Solutions

In the previous section, we noticed that the existence of classical solution (more precisely, of solutions $\mu \in \mathcal{M}_2^+(\Omega)$) may fail when $A$ and $B$ are convex; on the other hand, we show now that when the latter functions are concave, this existence is guaranteed. Indeed, we will prove Theorem 4.26, which provides the existence of optimal measures in $\mathcal{M}_2^+(\Omega)$ when the function $D$ defined in (2.29) is concave in the first variable. In particular, in Lemma 4.25 below we show that the latter concavity condition holds, for instance, when both $A$ and $B$ are concave. In the following claim, we denote by $g'_\pm(s)$ the left and right derivative of any function $g : \mathbb{R} \to \mathbb{R}$ at $s$, which always exist when $g$ is concave.

**Lemma 4.25.** *If both $A$ and $B$ are concave, then the function $D$ defined in (2.29) is concave; moreover, $D$ is strictly concave if, in addition, either $A$ or $B$ is strictly concave and*

$$B'_+(0) < A'_-(\operatorname{diam} \Omega)\,.$$

*Proof.* Assume that both $A$ and $B$ are concave. Then so is the function

$$l \mapsto A\left(\frac{a+\tilde{a}}{2} + l\right) + B\left(b - \frac{a+\tilde{a}}{2} - l\right). \tag{4.52}$$

As a consequence, the infimum of this function in the interval

$$0 \le l \le b - \frac{a+\tilde{a}}{2}$$

(which equals $D\big((a+\tilde{a})/2, b\big)$ by definition) is a minimum, and it is attained either at $l = 0$ or at $l = b - (a+\tilde{a})/2$. In the first case, we have

$$\begin{aligned} D\left(\frac{a+\tilde{a}}{2}, b\right) &= A\left(\frac{a+\tilde{a}}{2}\right) + B\left(b - \frac{a+\tilde{a}}{2}\right) \\ &\ge \frac{A(a) + A(\tilde{a}) + B(b-a) + B(b-\tilde{a})}{2} \\ &\ge \frac{D(a,b) + D(\tilde{a},b)}{2}\,. \end{aligned} \tag{4.53}$$

On the other hand, in the second case one has

$$D\left(\frac{a+\tilde{a}}{2},b\right)=A(b)+B(0)\geq D(a,b)\vee D(\tilde{a},b)\geq\frac{D(a,b)+D(\tilde{a},b)}{2}\,. \tag{4.54}$$

Hence, we have proved the concavity of $D(\cdot,b)$, minding that it is a continuous function.

Concerning the strict concavity, notice that the first inequality in (4.53) is strict if either $A$ or $B$ is strictly concave; but the inequality (4.54) may fail to be strict even if both $A$ and $B$ are strictly concave. For instance, if $B'(s) > A'(l)$ for any $s$ and $l$, then $D(\alpha,b)=A(b)+B(0)$ for any $\alpha$, and then the concavity of $D(\cdot,b)$ is not strict. On the other hand, if $B'_+(0)<A'_-(\operatorname{diam}\Omega)$, then the infimum in (4.52) is always attained at $l=0$: the meaning of this fact is that, if the ticket is sufficiently cheap, then for each path it is convenient to use as much as possible of the public transportation network, or in other words $\delta_\mu=\bar{\delta}_\mu$ for each measure $\mu$. Since the inequality (4.53) is strict if either $A$ or $B$ is strictly concave (both being concave) then $D(\cdot,b)$ is strictly concave and the proof is achieved. □

**Theorem 4.26.** *If the function $D(\cdot,b)$ is concave (resp. strictly concave) for any $b\in\mathbb{R}^+$, then there exists an optimal measure $\mu$ in $\mathcal{M}_2^+(\Omega)$ (resp. any optimal measure $\mu\in\mathcal{M}_1^+(\Omega)$ belongs to $\mathcal{M}_2^+(\Omega)$). In particular, by Lemma 4.25, there is an optimal measure $\mu\in\mathcal{M}_2^+(\Omega)$ if both $A$ and $B$ are concave; moreover, each optimal measure $\mu\in\mathcal{M}_1^+(\Omega)$ belongs to $\mathcal{M}_2^+(\Omega)$ if both $A$ and $B$ are concave, at least one of them is strictly concave and, in addition, $B'_+(0)<A'_-(\operatorname{diam}\Omega)$.*

*Proof.* Assume the existence of an optimal measure

$$a\mathscr{H}^1\llcorner\Sigma=\mu\in\mathcal{M}_1^+(\Omega)\setminus\mathcal{M}_2^+(\Omega)\,,$$

where, as usual, $\Sigma$ is a rectifiable set and $a:\Sigma\to[0,1]$ a Borel function; since $\mu\notin\mathcal{M}_2^+(\Omega)$ we have that $0<a(x)<1$ in a set of strictly positive length. As an immediate consequence, we can find $\varepsilon>0$ and two sets $\Sigma_1$, $\Sigma_2$ such that

$$\begin{aligned}&\mu(\Sigma_1)=\mu(\Sigma_2)>0\,, && \Sigma_1\cap\Sigma_2=\emptyset\,,\\ &\varepsilon<a(x)<1-\varepsilon && \text{for any } x\in\Sigma_1\cup\Sigma_2\,.\end{aligned}$$

We define therefore

$$\mu_s:=\mu+s\big(\mu\llcorner\Sigma_1-\mu\llcorner\Sigma_2\big)\,;$$

by construction, for any $-\varepsilon<s<\varepsilon$ one has $\mu_s\in\mathcal{M}_1^+(\Omega)$ and $\|\mu_s\|=\|\mu\|$. Take now any $\theta\in\Theta$: since $s\mapsto\mu_s(\theta)$ is a linear map, then

$$s\mapsto D\big(\mathscr{H}^1(\theta)-\mu_s(\theta),\mathscr{H}^1(\theta)\big)$$

is concave or strictly concave if so is $D$ in the first variable. By the definition (4.32) of $\bar{\delta}_\mu$, the equality

$$\bar{\delta}_\mu(\theta) = D\big(\mathscr{H}^1(\theta) - \mu(\theta), \mathscr{H}^1(\theta)\big)$$

holds; in particular, it can be obtained in the same way as formula (2.30). Therefore, for a generic path $\theta \in \Theta$ the map $s \mapsto \bar{\delta}_{\mu_s}(\theta)$ is concave or strictly concave. Take now an optimal t.p.m. $\eta$ for the cost $\overline{C}_\mu$, and observe that

$$\text{the map } s \mapsto \overline{C}_{\mu_s}(\eta) \text{ is concave (or strictly concave) if so is } D(\cdot, b)\,. \tag{4.55}$$

We now consider the strictly concave case: by (4.55) and the fact that $\mu_s$ is defined for $s$ in a neighborhood of zero we derive the existence of some $-\varepsilon < \bar{s} < \varepsilon$ such that $\overline{C}_{\mu_{\bar{s}}}(\eta) < \overline{C}_\mu(\eta)$. Since $\eta$ is an optimal t.p.m. for $\overline{C}_\mu$, it follows

$$MK(\mu_{\bar{s}}) \leq \overline{C}_{\mu_{\bar{s}}}(\eta) < \overline{C}_\mu(\eta) = MK(\mu)\,,$$

and since $\|\mu_{\bar{s}}\| = \|\mu\|$,

$$H(\|\mu_{\bar{s}}\| = H(\|\mu\|)$$

and hence $\mathfrak{F}(\mu_{\bar{s}}) < \mathfrak{F}(\mu)$ against the optimality of $\mu$. This contradiction shows that *any optimal measure* $\mu \in \mathcal{M}_1^+(\Omega)$ belongs to $\mathcal{M}_2^+(\Omega)$ in the strictly concave case (in particular, this means that any optimal measure $\mu$ belongs to $\mathcal{M}_2^+(\Omega)$ under the assumptions of Proposition 4.20).

Let us consider now the concave case: it is still true by (4.55) that $s \mapsto \overline{C}_{\mu_s}(\eta)$ is concave for $-\varepsilon < s < \varepsilon$; hence, for any given optimal measure $\mu$, either there exists some $\bar{s}$ such that $\overline{C}_{\mu_{\bar{s}}}(\eta) < \overline{C}_\mu(\eta)$, and then we conclude as in the previous case, or one has

$$\overline{C}_{\mu_s}(\eta) = \overline{C}_\mu(\eta) \qquad \text{for each } -\varepsilon < s < \varepsilon\,. \tag{4.56}$$

We denote by $\mathcal{N}$ the set of all the optimal measures belonging to $\mathcal{M}_1^+(\Omega)$, and consider the auxiliary problem of minimizing, over $\mathcal{N}$, the functional $\widetilde{\mathfrak{F}}$ corresponding to the choice

$$\widetilde{A}(t) := \sqrt{t}\,, \qquad \widetilde{B}(t) := 0\,.$$

Since $\mathfrak{F}$ is l.s.c. and coercive by construction, the set $\mathcal{N}$ is a bounded and weakly* closed subset of $\mathcal{M}^+(\Omega)$, hence it is also sequentially compact with respect to the weak* convergence. As a consequence, since $\widetilde{\mathfrak{F}}$ is also l.s.c. and coercive, there is a measure $\mu$ minimizing $\widetilde{\mathfrak{F}}$ over $\mathcal{N}$.

As noticed before, since $\mu$ minimizes $\mathfrak{F}$ then (4.56) holds; thus all the measures $\mu_s$ belong to $\mathcal{N}$ for $-\varepsilon < s < \varepsilon$. By Lemma 4.25, $\widetilde{D}$ is strictly concave in its first variable; therefore, the above arguments give the analogue of (4.55), so that the map $s \mapsto \widetilde{\mathfrak{F}}(\mu_s)$ is strictly concave: hence, there is some $-\varepsilon < \bar{s} < \varepsilon$ such that $\widetilde{\mathfrak{F}}(\mu_{\bar{s}}) < \widetilde{\mathfrak{F}}(\mu)$. Since $\mu_{\bar{s}} \in \mathcal{N}$, this gives a contradiction with the fact that $\mu$ minimizes $\widetilde{\mathfrak{F}}$ over $\mathcal{N}$. So, we deduce the existence of *some optimal measure* $\mu$ belonging to $\mathcal{M}_2^+(\Omega)$. □

# Chapter 5
# Topological Properties of Optimal Sets

In Chapter 4 we proved the existence of relaxed solutions to our network optimization problem, and gave conditions on the existence of classical solutions. Under such conditions, it becomes important to study some qualitative properties of an optimal set $\Sigma$, and in particular its closedness and connectedness.

For this purpose we introduce the idea of "transiting mass", that is the total mass passing through a single point. Even though the case we will mainly concentrate our attention on is when $A(t) = t$ and $B(t) = 0$, for the sake of generality we first present this definition in a more general framework.

## 5.1 Transiting Mass Function

In order to define and study the properties of the "transiting mass", we need to introduce a suitable set $\Theta_\mu$ of paths.

**Definition 5.1.** For every measure $\mu \in \mathcal{M}_1^+(\Omega)$, we define $\Theta_\mu$ as the subset of $\Theta$ made by all the paths $\theta \in \Theta$ satisfying

$$J\big(\mathscr{H}^1(\theta) - \mu(\theta), \mu(\theta)\big) = A\big(\mathscr{H}^1(\theta) - \mu(\theta)\big) + B\big(\mu(\theta)\big)\,,$$

where $J$ is given by (2.24).

In other words, $\Theta_\mu$ is the set of those paths for which $\delta_\mu(\theta) = \bar{\delta}_\mu(\theta)$, i.e. the most convenient choice for a passenger traveling along $\theta$ is to use the network as much as possible. Note that $\Theta_\mu = \Theta$ if $A(t) = t$ and $B(t) = 0$, or more generally if $A \geq B$.

**Definition 5.2.** Assume that $A$ and $B$ are $C^1$ functions and let $\eta$ be a t.p.m., the support of which is made of paths of uniformly bounded lengths, and minimizing the cost $\overline{C}_\mu$. We define the *generalized transiting mass* as the function $\alpha_\eta : \Omega \to \mathbb{R}^+$ given by

G. Buttazzo et al., *Optimal Urban Networks via Mass Transportation*,
Lecture Notes in Mathematics 1961, DOI: 10.1007/978-3-540-85799-0_5,

$$\alpha_\eta(x) := \int_{\Theta_\mu} \Big( A'\big(\mathscr{H}^1(\theta) - \mu(\theta)\big) - B'\big(\mu(\theta)\big) \Big)^+ \chi_\theta(x)\, d\eta(\theta)\,.$$

In particular, when $A(t) = t$ and $B(t) = 0$, we have

$$\alpha_\eta(x) = \eta\{\theta \in \Theta : \ x \in \theta\}\,, \tag{5.1}$$

which justifies the name of transiting mass function.

*Remark 5.3.* Note that the function

$$v(\theta) := A'\big(\mathscr{H}^1(\theta) - \mu(\theta)\big) - B'\big(\mu(\theta)\big)$$

is non-negative for any $\theta \in \Theta_\mu$ such that $\mu(\theta) > 0$. Indeed, if $\mu(\theta) > 0$ and $v(\theta) < 0$ then the minimum in the definition (2.24) is not achieved at $l = \mu(\theta)$, contradicting the assumption $\theta \in \Theta_\mu$.

We now show the upper semicontinuity of $\alpha_\eta$ and the fact that every optimal $\Sigma$ must be an upper level of $\alpha_\eta$.

**Lemma 5.4.** *The function $\alpha_\eta$ is u.s.c.*

*Proof.* We write

$$\alpha_\eta(x) = \int_{\Theta_\mu} \chi_\theta(x) v^+(\theta)\, d\eta(\theta)\,,$$

and the thesis follows by Fatou's Lemma. Indeed,

$$\chi_\theta(x) \geq \limsup \chi_\theta(x_n)$$

whenever $x_n \to x$, since a path $\theta$ containing infinitely many $x_n$ must also contain their limit $x$. Moreover, $\theta \mapsto v^+(\theta)$ is a positive Borel function uniformly bounded from above since $A$ and $B$ are of class $\mathrm{C}^1$ and $\mathscr{H}^1(\theta)$ is uniformly bounded over $\operatorname{spt}\eta$ by assumption. □

**Theorem 5.5.** *Let $\mu \in \mathcal{M}_1^+(\Omega)$ be an optimal measure and let $\eta \in \mathcal{M}^+_{\mu,0}(\Theta)$ be an optimal t.p.m., the support of which is made of paths of uniformly bounded lengths. Then $\mu$ is an upper level of $\alpha_\eta$, in the sense that*

$$\mathscr{H}^1 \llcorner \big\{\alpha_\eta(x) > r\big\} \leq \mu \leq \mathscr{H}^1 \llcorner \big\{\alpha_\eta(x) \geq r\big\}\,. \tag{5.2}$$

*Proof.* We start defining

$$\mu_\varepsilon := \mu \vee (\mathscr{H}^1 \llcorner \Sigma_\varepsilon)$$

for a small number $\varepsilon$, where $\Sigma_\varepsilon$ is a Borel set chosen in such a way that

$$\|\mu_\varepsilon - \mu\| = \varepsilon\,.$$

In order to evaluate $C_{\mu_\varepsilon}(\eta)$, we notice that for any $\theta \notin \Theta_\mu$, one has by definition $\bar{\delta}_{\mu_\varepsilon}(\theta) = \bar{\delta}_\mu(\theta)$ for $\varepsilon \le \bar{\varepsilon}(\theta)$. Defining then

$$\Theta'_\varepsilon := \{\theta \in \Theta \setminus \Theta_\mu : \varepsilon \le \bar{\varepsilon}(\theta)\}\,, \qquad \Theta''_\varepsilon := (\Theta \setminus \Theta_\mu) \setminus \Theta'_\varepsilon\,,$$

and recalling that $\mathscr{H}^1(\theta)$ is bounded for $\theta \in \operatorname{spt}\eta$, one finds

$$\begin{aligned} &\bar{\delta}_{\mu_\varepsilon}(\theta) = \bar{\delta}_\mu(\theta) \quad \text{on } \Theta'_\varepsilon\,, \qquad \bar{\delta}_{\mu_\varepsilon}(\theta) = \bar{\delta}_\mu(\theta) + O(\varepsilon) \quad \text{on } \Theta''_\varepsilon\,, \\ &\eta(\Theta''_\varepsilon) = o(1)\,, \end{aligned} \tag{5.3}$$

the second equality being true due to the fact that $A, B \in C^1$, and the third equality being true due to the fact that $\Theta''_\varepsilon \searrow \emptyset$ as $\varepsilon \to 0$ (all the infinitesimals here and below are to be intended for $\varepsilon \to 0$, and are always uniform with respect to $\theta$). On the other hand, we claim that

$$\int_{\Theta_\mu} \big(\bar{\delta}_{\mu_\varepsilon}(\theta) - \bar{\delta}_\mu(\theta)\big)\, d\eta(\theta) = \int_{\Theta_\mu} -v^+(\theta)\big(\mu_\varepsilon(\theta) - \mu(\theta)\big)\, d\eta(\theta) + o(\varepsilon)\,. \tag{5.4}$$

Indeed, by definition of $\bar{\delta}_\mu$ we easily obtain that

$$\bar{\delta}_\mu(\theta) = A\big(\mathscr{H}^1(\theta) - \mu(\theta)\big) + B\big(\mu(\theta)\big)\,,$$

and that

$$\bar{\delta}_{\mu_\varepsilon}(\theta) = A\big(\mathscr{H}^1(\theta) - \mu(\theta) - \lambda(\theta)\big) + B\big(\mu(\theta) + \lambda(\theta)\big)$$

for some

$$0 \le \lambda(\theta) \le \mu_\varepsilon(\theta) - \mu(\theta) \le \varepsilon\,.$$

Thus, recalling that $A$ and $B$ are $\mathrm{C}^1$ functions, we get

$$\bar{\delta}_{\mu_\varepsilon}(\theta) - \bar{\delta}_\mu(\theta) = -\lambda(\theta)\,\big(v(\theta) + o(1)\big) \tag{5.5}$$

as $\varepsilon \to 0^+$, the infinitesimal in the above formula being uniform with respect to $\theta$ because the function $\theta \mapsto \mathscr{H}^1(\theta)$ is uniformly bounded over $\operatorname{spt}\eta$ by Lemma 4.6 and $\mu(\theta) \le \|\mu\|$. Notice now that if $v(\theta) > 0$ then

$$\lambda(\theta) = \mu_\varepsilon(\theta) - \mu(\theta)$$

for $\varepsilon$ sufficiently small depending on $\theta$; analogously, if $v(\theta) < 0$, then $\lambda(\theta) = 0$ for $\varepsilon$ small enough. Hence, the two integrands in formula (5.4) coincide at $\theta$ whenever $\varepsilon$ is small enough (depending on $\theta$), and are of order $O(\varepsilon)$ uniformly with respect to $\theta$ due to (5.5) and to the fact that $v(\theta)$ is bounded. Therefore, (5.4) follows.

Making use of (5.3) and (5.4), we can now evaluate

$$
\begin{aligned}
\overline{C}_{\mu_\varepsilon}(\eta) - \overline{C}_\mu(\eta) &= \int_\Theta \left(\bar{\delta}_{\mu_\varepsilon}(\theta) - \bar{\delta}_\mu(\theta)\right) d\eta(\theta) \\
&= \int_{\Theta_\mu} \left(\bar{\delta}_{\mu_\varepsilon}(\theta) - \bar{\delta}_\mu(\theta)\right) d\eta(\theta) + \int_{\Theta\setminus\Theta_\mu} \left(\bar{\delta}_{\mu_\varepsilon}(\theta) - \bar{\delta}_\mu(\theta)\right) d\eta(\theta) \\
&= \int_{\Theta_\mu} -v^+(\theta)\big(\mu_\varepsilon(\theta) - \mu(\theta)\big)\, d\eta(\theta) + o(\varepsilon) + \int_{\Theta''_\varepsilon} \left(\bar{\delta}_{\mu_\varepsilon}(\theta) - \bar{\delta}_\mu(\theta)\right) d\eta(\theta) \\
&= \int_{\Theta_\mu} -v^+(\theta)\big(\mu_\varepsilon(\theta) - \mu(\theta)\big)\, d\eta(\theta) + o(\varepsilon) \\
&= \int_{\Theta_\mu} -v^+(\theta)\Big(\int_{\Sigma_\varepsilon} \chi_\theta(x)\, d(\mu_\varepsilon - \mu)(x)\Big)\, d\eta(\theta) + o(\varepsilon) \\
&= \int_{\Sigma_\varepsilon}\int_{\Theta_\mu} -v^+(\theta)\chi_\theta(x)\, d\eta(\theta)\, d(\mu_\varepsilon - \mu)(x) + o(\varepsilon) \\
&= -\int_{\Sigma_\varepsilon} \alpha_\eta(\theta)\, d(\mu_\varepsilon - \mu)(x) + o(\varepsilon)\,.
\end{aligned}
$$

Analogously, we define

$$\mu'_\varepsilon := \mu - (\mu \llcorner \Sigma'_\varepsilon)$$

for a set $\Sigma'_\varepsilon$ such that $\mu(\Sigma'_\varepsilon) = \varepsilon$. A similar computation as above allows to evaluate

$$\overline{C}_{\mu'_\varepsilon}(\eta) - \overline{C}_\mu(\eta) = \int_{\Sigma'_\varepsilon} \alpha_\eta(x)\, d(\mu - \mu'_\varepsilon)(x) + o(\varepsilon)\,. \tag{5.6}$$

Finally, we set

$$\mu''_\varepsilon = \mu \vee \mathscr{H}^1 \llcorner \Sigma_\varepsilon - \mu \llcorner \Sigma'_\varepsilon\,,$$

where $\Sigma_\varepsilon$ and $\Sigma'_\varepsilon$ are two sets as before and having empty intersection. We deduce

$$C_{\mu''_\varepsilon}(\eta) - C_\mu(\eta) = -\int_{\Sigma_\varepsilon} \alpha_\eta(x)\, d(\mu_\varepsilon - \mu)(x) + \int_{\Sigma'_\varepsilon} \alpha_\eta(x)\, d(\mu - \mu'_\varepsilon)(x) + o(\varepsilon)\,.$$

where

$$\mu_\varepsilon := \mu \vee \mathscr{H}^1 \llcorner \Sigma_\varepsilon\,, \qquad \mu'_\varepsilon := \mu - (\mu \llcorner \Sigma'_\varepsilon)\,.$$

We conclude by pointing out that, if $\Sigma$ does not satisfy (5.2), then there exist two numbers $r > s \geq 0$, and two sets

$$\Sigma_\varepsilon \subseteq \{x \,:\, \alpha_\eta(x) \geq r\}\,, \qquad \Sigma'_\varepsilon \subseteq \{x \,:\, \alpha_\eta(x) \leq s\}$$

with $\mu(\Sigma'_\varepsilon) = \varepsilon$ and

$$\mu \llcorner \Sigma_\varepsilon \leq \mathscr{H}^1 \llcorner \Sigma_\varepsilon\,, \qquad \big\|\mathscr{H}^1 \llcorner \Sigma_\varepsilon - \mu \llcorner \Sigma_\varepsilon\big\| = \varepsilon$$

for $\varepsilon > 0$ small enough. Hence, with the above notation

$$-\int_{\Sigma_\varepsilon} \alpha_\eta(x)\, d(\mu_\varepsilon - \mu)(x) + \int_{\Sigma'_\varepsilon} \alpha_\eta(x)\, d(\mu - \mu'_\varepsilon)(x) \le -\beta\varepsilon$$

for

$$\beta := \frac{r-s}{2} > 0$$

and for all $\varepsilon > 0$ small enough. Since $\|\mu\| = \|\mu''_\varepsilon\|$ and by the fact that $\eta$ is an optimal t.p.m. for $\mu$, this gives $\mathfrak{F}(\mu''_\varepsilon) < \mathfrak{F}(\mu)$ for $\varepsilon$ small enough, against the optimality of $\mu$. □

*Remark 5.6.* It is clear that among the numbers $r \ge 0$ satisfying (5.2) there is a maximum one.

**Lemma 5.7.** *Given an optimal measure $\mu$ and an optimal t.p.m. $\eta$, for any $r \ge 0$ such that $\mu \ge \mathscr{H}^1 \llcorner S$, where*

$$S := \left\{\alpha_\eta(x) > r\right\},$$

*one has that $S$ is contained in countably many geodesics $\theta \in \operatorname{spt}\eta$ (up to an $\mathscr{H}^1$−negligible set).*

*Proof.* Consider first the case $r > 0$. Take $\theta_1$ which "almost maximizes"

$$\mathscr{H}^1(\theta \cap S)$$

among all paths $\theta \in \operatorname{spt}\eta$ in the sense that

$$\mathscr{H}^1(\theta \cap S) \le 2\mathscr{H}^1(\theta_1 \cap S) \qquad \forall\, \theta \in \operatorname{spt}\eta\,.$$

Let us now define, inductively, $\theta_{h+1} \in \operatorname{spt}\eta$ as a path almost maximizing

$$\mathscr{H}^1\left(\theta \cap \left(S \setminus \bigcup_{j=1}^h \theta_j\right)\right)$$

in $\operatorname{spt}\eta$. We claim that $S$ is $\mathscr{H}^1$−essentially contained in $\cup_{h\in\mathbb{N}}\theta_h$, that is,

$$\mathscr{H}^1(S \setminus S_0) = 0\,, \qquad \text{where} \qquad S_0 := S \cap \left(\cup_{h\in\mathbb{N}}\, \theta_h\right).$$

Indeed, since $\alpha_\eta \ge r > 0$ inside $S$, assuming $\mathscr{H}^1(S \setminus S_0) > 0$ one would have

$$\begin{aligned}0 &< \int_{S\setminus S_0} \alpha_\eta(x)\, d\mathscr{H}^1(x) = \int_{S\setminus S_0}\int_{\Theta_\mu} \chi_\theta(x) v^+(\theta)\, d\eta(\theta)\, d\mathscr{H}^1(x)\\ &= \int_{\Theta_\mu}\int_{S\setminus S_0} \chi_\theta(x) v^+(\theta)\, d\mathscr{H}^1(x)\, d\eta(\theta)\,.\end{aligned}$$

Therefore, there exists $\bar\theta \in \operatorname{spt}\eta$ such that

$$\int_{S\setminus S_0} \chi_\theta(x) v^+(\bar\theta)\, d\mathscr{H}^1(x) > 0\,,$$

and hence

$$\tau := \mathscr{H}^1(\theta \cap (S \setminus S_0)) > 0\,.$$

By construction, $\theta_h \neq \bar{\theta}$ for all $h \in \mathbb{N}$, and consequently for any $h \in \mathbb{N}$

$$\mathscr{H}^1\left(\theta_{h+1} \cap \left(S \setminus \bigcup_{j=1}^{h} \theta_j\right)\right) > \frac{\tau}{2}\,.$$

This would give $\mathscr{H}^1(S) = \infty$, which is a contradiction since by hypothesis $\mathscr{H}^1(S) \leq \|\mu\|$, thus the proof is achieved for the case $r > 0$.

If $r = 0$, note that $S := \cup_k S_k$, where

$$S_k := \left\{\alpha_\eta(x) > \frac{1}{k}\right\}$$

and $\mu \geq \mathscr{H}^1 \llcorner S$ implies $\mu \geq \mathscr{H}^1 \llcorner S_k$ for each $k \in \mathbb{N}$. Hence, as proved above, each $S_k$ is contained in a countable number of geodesics of $\operatorname{spt} \eta$, and hence so is $S$. □

From now on, we will consider a particular situation, i.e.

$$A(t) = t\,, \qquad\qquad B(t) = 0\,. \tag{5.7}$$

This is the simplest case, and it has been already considered in many works with the further assumption that $\Sigma$ is connected (see e.g. [19, 20]). By the convexity of $A$ and $B$ and by Theorem 4.26, we already know that there exists an optimal measure $\mu = \mathscr{H}^1 \llcorner \Sigma \in \mathcal{M}_2^+(\Omega)$, even though there may be also optimal measures not contained in $\mathcal{M}_2^+(\Omega)$; indeed, one has $D(a, b) = a$, so that $D(\cdot, b)$ is concave but not strictly concave. Moreover, one has clearly $\bar{\delta}_\mu = \delta_\mu$ for any measure $\mu$ by definition of $\bar{\delta}_\mu$ given in (4.32). Hence by Lemma 4.17-ii) we infer the existence of optimal t.p.m.'s for $C_\mu$, and not just for $\overline{C}_\mu$; more precisely, (5.7) implies that $C_\mu = \overline{C}_\mu$.

We underline that, since $A$ and $B$ are now fixed, the hypotheses of the following results will be only about $H$. Notice that until now, we have not yet given any hypothesis on $H$. This is due to the fact that, in all the proofs of the previous chapter, starting from an optimal measure $\mu$ we have always built a "competitor" $\mu'$ having the same total mass: therefore, since

$$H(\|\mu\|) = H(\|\mu'\|)\,,$$

the function $H$ played no role. On the other hand, in the rest of the monograph we will often consider competitors having a different total mass, so that the hypotheses on $H$ will become of primary importance.

Note that the hypothesis $B \equiv 0$ means that "moving by train is free of charge". Then, for people who need to move from/to places not so far from the network, the most convenient thing should be to reach the network, take a train until they arrive close to their destination, then reach the latter by own means.

We stress that, since we are assuming (5.7), one has $\Theta_\mu = \Theta$ for any $\mu$, which makes the Definition 5.2 of the transiting mass more significant; in particular, as already remarked, $\alpha_\eta$ is given by

$$\alpha_\eta(x) = \eta\{\theta \in \Theta : \ x \in \theta\}\,.$$

Therefore, we can give the following useful calculation for $C_\mu(\eta) = \overline{C}_\mu(\eta)$, which shows that $\mu \mapsto C_\mu(\eta)$ is a constant plus the integral of a given function in $d\mu$.

$$\begin{aligned}
C_\mu(\eta) &= \int_\Theta \Big(\mathscr{H}^1(\theta) - \mu(\theta)\Big)\, d\eta(\theta) = C_\emptyset(\eta) - \int_\Theta \mu(\theta)\, d\eta(\theta) \\
&= C_\emptyset(\eta) - \int_\Theta \Big(\int_\Omega \chi_\theta(x))\, d\mu(x)\Big)\, d\eta(\theta) \\
&= C_\emptyset(\eta) - \int_\Omega \Big(\int_\Theta \chi_\theta(x)\, d\eta(\theta)\Big)\, d\mu(x) \\
&= C_\emptyset(\eta) - \int_\Omega \alpha_\eta(x)\, d\mu(x)\,.
\end{aligned} \tag{5.8}$$

We conclude this chapter with a couple of remarks: the first one is a consequence of Lemma 5.7 and of the above calculation of $C_\mu(\eta)$.

*Remark 5.8.* Let $\mu$ be an optimal measure and $\eta$ be an optimal t.p.m. As an immediate consequence of Lemma 5.7, we obtain that the measure

$$\mu' = \mu \llcorner \{\alpha_\eta(x) > 0\}$$

is concentrated on countably many geodesics $\theta \in \operatorname{spt}\eta$. By (5.8), we obtain that

$$C_{\mu'}(\eta) = C_\mu(\eta) + \int_\Omega \alpha_\eta(x)\, d(\mu - \mu')(x) = C_\mu(\eta)\,.$$

Hence, minding that $\mu' \leq \mu$, and thus $H(\|\mu'\|) \leq H(\|\mu\|)$, one has that also $\mu'$ is an optimal measure and that $\eta$ is an optimal t.p.m. also with respect to $\mu'$. As discussed in Proposition 4.20, we obtain that $\mu' = \mu$ whenever either (4.37) or (4.38) occurs. Summarizing, in either of these cases we have obtained that *any* optimal measure $\mu$ is concentrated on countably many geodesics contained in the support of an optimal t.p.m.; otherwise, an optimal measure $\mu' < \mu$ is concentrated on countably many geodesics, and the remaining part $\mu - \mu'$ is completely useless.

*Remark 5.9.* It is worth noticing that, if one is given a t.p.m. $\eta$ which is known to be optimal with respect to some unknown optimal measure, there is an easy method to determine this measure, or more precisely to determine all the measures $\mu$ which are optimal and with respect to which $\eta$ is an optimal t.p.m. Indeed, for any $l \geq 0$ define $\mathcal{M}_l$ to be the subset of $\mathcal{M}_1^+(\Omega)$ made by all those measures which are upper levels in the sense of (5.2) and with total mass equal to $l$. An immediate consequence of (5.8) and of the formula

$$\int_{\Omega} \alpha_\eta \, d\mu = \int_0^{+\infty} \mu(\{x : \alpha_\eta(x) \geq t\}) \, dt$$

is that for $\mu \in \mathcal{M}_l$ one has

$$C_\mu(\eta) = C_\emptyset(\eta) - \beta(l)$$

for a function $\beta(l)$ which depends on $l$ but not on $\mu$: more precisely, $\beta$ is given by the following formula (to be intended in the weak sense, or in the classical sense up to countably many points)

$$\begin{cases} \beta(0) = 0\,, \\ \beta'(l) = \inf\left\{t : \mathscr{H}^1\left(\{x : \alpha_\eta(x) \geq t\}\right) \leq l\right\}. \end{cases} \tag{5.9}$$

Arguing as in Theorem 5.5, we know that for any measure $\mu \in \mathcal{M}_1^+(\Omega)$ with $\|\mu\| = l$ one has

$$C_\mu(\eta) \geq C_\emptyset(\eta) - \beta(l)\,,$$

and the inequality is strict if and only if $\mu \notin \mathcal{M}_l$. Notice also that the function $l \mapsto \beta(l)$ is concave thanks to (5.9); more precisely, the left and the right derivative of $\beta$ at $l$ coincide respectively with the maximum and minimum numbers $r^+(l)$ and $r^-(l)$ such that (5.2) holds for a measure $\mu \in \mathcal{M}_l$: hence, $r^\pm(l)$ depend on $l$ but not on the choice of $\mu \in \mathcal{M}_l$, and $r^-(l_1) \geq r^+(l_2)$ whenever $l_1 < l_2$.

It is now clear that for any $\mu \in \mathcal{M}_l$ one has

$$\mathfrak{F}(\mu) \leq C_\emptyset(\eta) - \beta(l) + H(l)\,,$$

and the inequality is in fact an equality if and only if $\eta$ is an optimal t.p.m. with respect to $\mu$. Consider now the function $l \mapsto H(l) - \beta(l)$: since $H$ and $\beta$ are positive functions, $H(0) = \beta(0) = 0$, $H(l) \to \infty$ when $l \to \infty$ while $\beta(l)$ is bounded by $C_\emptyset(\eta)$ thanks to (5.8), we deduce that $\operatorname{Arg\,min}(H - \beta)$ is a non-empty compact subset $L$ of $\mathbb{R}^+$. Since we know that $\eta$ is an optimal t.p.m. for some optimal measure $\mu$, we finally conclude that the optimal measures for which $\eta$ is an optimal t.p.m. are precisely the measures belonging to some $\mathcal{M}_l$ with $l \in L$.

## 5.2 Ordered Transport Path Measures

In this section, we introduce the notion of *ordered* transport path measures, which are t.p.m. with additional useful properties. First of all, we select an optimal measure $\mu \in \mathcal{M}_1^+(\Omega)$; then, we notice that by the lower semicontinuity of $\mathfrak{F}$ the set $\mathcal{M}_{\mu,0}^+(\Theta)$ of optimal t.p.m.'s is a non-empty, weakly$^*$ closed and convex subset of $\mathcal{M}^+(\Theta)$. We define then the following two sets.

**Definition 5.10.** Let

$$\mathcal{M}^+_{\mu,1}(\Theta) := \operatorname{Arg\,min}\Big\{ \int_\Theta \ell(\theta)\, d\eta(\theta) : \ \eta \in \mathcal{M}^+_{\mu,0}(\Theta) \Big\}, \tag{5.10}$$

$$\mathcal{M}^+_{\mu,2}(\Theta) := \operatorname{Arg\,max}\Big\{ \int_\Sigma \alpha_\eta(x)^2\, d\mathscr{H}^1(x) : \ \eta \in \mathcal{M}^+_{\mu,1}(\Theta) \Big\}, \tag{5.11}$$

where

$$\ell(\theta) := \int_0^1 |\theta'(t)|\, dt$$

is the *parametric length* of $\theta$. The elements of $\mathcal{M}^+_{\mu,2}(\Theta)$ will be called *ordered transport path measures.*

*Remark 5.11.* Notice that, similarly to $\mathcal{M}^+_{\mu,0}(\Theta)$, also $\mathcal{M}^+_{\mu,1}(\Theta)$ and $\mathcal{M}^+_{\mu,2}(\Theta)$ are non-empty and weakly* closed subsets of $\mathcal{M}^+(\Theta)$. In fact, $\theta \mapsto \ell(\theta)$ is l.s.c., hence the map $\eta \mapsto \int_\Theta \ell(\theta)\, d\eta$ is l.s.c., which implies the claim concerning $\mathcal{M}^+_{\mu,1}(\Theta)$. Moreover, the map $\eta \to \alpha_\eta(x)$ is u.s.c. for any $x \in \Sigma$ by (5.1) since the set

$$\{\theta \in \Theta : \ x \in \theta\}$$

is closed, while

$$\alpha_\eta(x) \le \eta(\Theta) = \|f^+\|,$$

and hence the fact that $\mathcal{M}^+_{\mu,2}(\Theta)$ is weakly* closed and non-empty follows with the help of Fatou's lemma. Moreover, clearly both $\mathcal{M}^+_{\mu,0}(\Theta)$ and $\mathcal{M}^+_{\mu,1}(\Theta)$ (but not $\mathcal{M}^+_{\mu,2}(\Theta)$) are convex sets.

As we will see in the sequel, the elements of $\mathcal{M}^+_{\mu,2}(\Theta)$ have rather particular properties among the optimal t.p.m.'s. For this reason, in what follows we will restrict our attention to the ordered t.p.m.'s.

We notice first an easy consequence of Lemma 4.1.

**Corollary 5.12.** *For any t.p.m.* $\eta \in \mathcal{M}^+_{\mu,0}(\Theta)$ *there is a path* $\sigma$ *maximizing*

$$\theta \mapsto \mu(\theta) = \mathscr{H}^1(\theta \cap \Sigma)$$

*within* $\operatorname{spt} \eta$.

*Proof.* Just take a maximizing sequence $\{\theta_n\}$, extract a subsequence converging to a path $\theta$ with respect to $d_\Theta$ (this is possible since all these paths have uniformly bounded Euclidean lengths), notice that $\theta \in \operatorname{spt}\eta$ because $\operatorname{spt}\eta$ is closed by definition, and finally apply Lemma 4.1 with $X := \Theta$, $C_n := \theta_n$ and $\nu := \mu$. □

We show now the following geometric property of the paths belonging to $\operatorname{spt}\eta$ with $\eta \in \mathcal{M}^+_{\mu,0}(\Theta)$.

**Lemma 5.13.** *Let $\eta \in \mathcal{M}^+_{\mu,0}(\Theta)$ and let $\theta \in \operatorname{spt}\eta$ be a path such that*

$$\mu(\theta([t_1, t_2])) = a > 0$$

*for some $0 \le t_1 < t_2 \le 1$. Then for any $\varepsilon > 0$ there is $\rho > 0$ such that the following holds: given any $\tau \in \operatorname{spt}\eta \cap B_\Theta(\theta, \rho)$ and considering on $\tau$ a parametrization such that $|\tau(t_i) - \theta(t_i)| \le \rho$ for $i = 1, 2$, one has*

$$\mu(\tau([t_1, t_2])) > a - \varepsilon .$$

In particular, the above statement with $t_1 = 0$ and $t_2 = 1$ implies the lower semicontinuity of $\theta \mapsto \mu(\theta)$ in $\operatorname{spt}\eta$. Since the upper semicontinuity holds in the whole $\Theta$ by Lemma 4.1, we derive that the quantity of network taken by the paths, that is $\theta \mapsto \mu(\theta)$, is a continuous function in $\operatorname{spt}\eta$.

*Proof (of Lemma 5.13).* By the lower semicontinuity of the Euclidean length, we know that

$$\mathscr{H}^1(\theta_{|[t_1,t_2]}) \le \mathscr{H}^1(\tau_{|[t_1,t_2]}) + \frac{\varepsilon}{2}, \tag{5.12}$$

provided that $\rho$ is sufficiently small. We claim that if the assertion is not true, that is, if

$$\mu(\tau([t_1, t_2])) \le a - \varepsilon , \tag{5.13}$$

then $\tau$ is not a geodesic between $\tau(0)$ and $\tau(1)$: since $\eta$ is an optimal t.p.m., this will give a contradiction with Lemma 4.17-iii), hence concluding the proof.

To show the claim, define $\tilde{\tau}$ as the path that equals $\tau$ outside of $[t_1, t_2]$ and that is defined as follows between $t_1$ and $t_2$: a segment connecting $\tau(t_1)$ to $\theta(t_1)$, then the path $\theta_{|[t_1,t_2]}$ between $\theta(t_1)$ and $\theta(t_2)$, then the segment connecting $\theta(t_2)$ to $\tau(t_2)$. Figure 5.1 shows the situation, where $\tau$ is the upper path, $\theta$ the lower one, and $\tilde{\tau}$ the dark shaded one. Then, minding that by (5.13) one has

$$\delta_\mu(\tilde{\tau}_{|[t_1,t_2]}) \ge \mathscr{H}^1(\tau_{|[t_1,t_2]}) - a + \varepsilon , \tag{5.14}$$

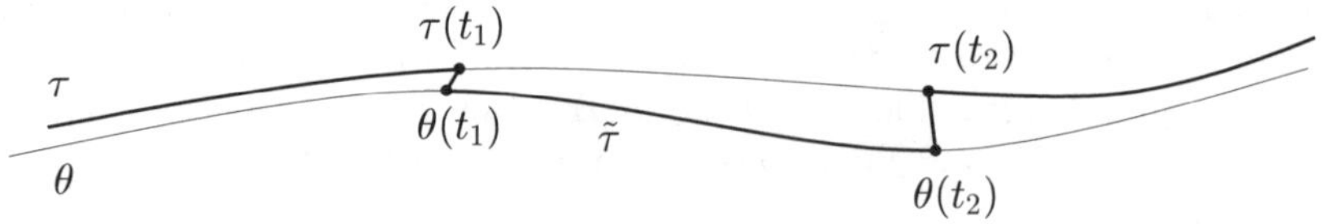


**Fig. 5.1** Situation in Lemma 5.13

we find

$$\begin{aligned}
\delta_\mu(\tilde\tau) &= \delta_\mu(\tilde\tau_{|[0,t_1]}) + \delta_\mu(\tilde\tau_{|[t_1,t_2]}) + \delta_\mu(\tilde\tau_{|[t_2,1]})\\
&\le \delta_\mu(\tau_{|[0,t_1]}) + 2\rho + \delta_\mu(\theta_{|[t_1,t_2]}) + \delta_\mu(\tau_{|[t_2,1]})\\
&= \delta_\mu(\tau_{|[0,t_1]}) + 2\rho + \mathscr{H}^1(\theta_{|[t_1,t_2]}) - a + \delta_\mu(\tau_{|[t_2,1]})\\
&\le \delta_\mu(\tau_{|[0,t_1]}) + 2\rho + \mathscr{H}^1(\tau_{|[t_1,t_2]}) + \varepsilon/2 - a + \delta_\mu(\tau_{|[t_2,1]}) \qquad \text{(by (5.12))}\\
&\le \delta_\mu(\tau_{|[0,t_1]}) + 2\rho + \delta_\mu(\tau_{|[t_1,t_2]}) + a - \varepsilon/2 - a + \delta_\mu(\tau_{|[t_2,1]}) \qquad \text{(by (5.14))}\\
&= \delta_\mu(\tau) + 2\rho - \varepsilon/2 < \delta_\mu(\tau),
\end{aligned}$$

provided $\rho < \varepsilon/4$, which, recalling that $\delta_\mu = \bar\delta_\mu$ thanks to (5.7), gives the desired contradiction. □

**Corollary 5.14.** *In the hypotheses of Lemma 5.13, for any $\varepsilon > 0$ there is a $\rho > 0$ such that for any $\tau \in \operatorname{spt}\eta \cap B_\Theta(\theta,\rho)$ one has*

$$\mu\big(\tau([t_1,t_2]) \cap \theta([t_1,t_2])\big) > a - \varepsilon\,.$$

*Proof.* Assume that the thesis is not true, i.e. there exists an $\varepsilon > 0$ such that for every $\rho > 0$ there is a $\tau_\rho \in \operatorname{spt}\eta \cap B_\Theta(\theta,\rho)$ with

$$\mu\big(\tau_\rho([t_1,t_2]) \cap \theta([t_1,t_2])\big) \le a - \varepsilon\,.$$

Then, by Lemma 5.13 then there exists a $\rho_1 > 0$ such that, taking

$$\tau_1 := \tau_{\rho_1} \in \operatorname{spt}\eta\,,$$

one has

$$\mu\big(\tau_1([t_1,t_2])\big) > a - \frac{\varepsilon}{2}\,,$$

so that

$$\mu\big(\tau_1([t_1,t_2]) \setminus \theta([t_1,t_2])\big) > \frac{\varepsilon}{2}\,.$$

Suppose now that for some $k \in \mathbb{N}$ and for each $n \le k$ we constructed a $\rho_n \le \rho_{n-1}$ and a path

$$\tau_n := \tau_{\rho_n} \in B_\Theta(\theta,\rho_{n-1}) \cap \operatorname{spt}\eta$$

such that

$$\mu\Big(\tau_n([t_1,t_2]) \setminus \big(\bigcup_{j=1}^{n-1} \tau_j([t_1,t_2]) \cup \theta([t_1,t_2])\big)\Big) > \frac{\varepsilon}{n+1}\,. \tag{5.15}$$

Mind that

$$\mu(\tau_\rho([t_1,t_2]) \setminus \theta([t_1,t_2])) \to 0 \qquad \text{as} \qquad \rho \to 0$$

since

$$\limsup_{\rho\to 0} \mu(\tau_\rho([t_1,t_2])) \le \mu(\theta([t_1,t_2]))$$

by Lemma 4.1. Therefore,

$$\mu\left((\tau_\rho([t_1,t_2]) \setminus \theta([t_1,t_2])) \cap \left(\bigcup_{j=1}^k \tau_j([t_1,t_2])\right)\right) \to 0$$

as $\rho \to 0$. Then there exists a $\rho_{k+1} \le \rho_k$ such that, taking $\tau_{k+1} := \tau_{\rho_{k+1}}$, one has

$$\mu\left((\tau_{k+1}([t_1,t_2]) \setminus \theta([t_1,t_2])) \cap \left(\bigcup_{j=1}^k \tau_j([t_1,t_2])\right)\right) \le \frac{\varepsilon}{2} - \frac{\varepsilon}{k+2},$$

while

$$\mu(\tau_{k+1}([t_1,t_2])) > a - \frac{\varepsilon}{2}$$

by Lemma 5.13, so that

$$\begin{aligned}\mu\Big(\tau_{k+1}([t_1,t_2]) \setminus \big(\bigcup_{j=1}^k \tau_j([t_1,t_2]) \cup \theta([t_1,t_2])\big)\Big) &= \\ \mu(\tau_{k+1}([t_1,t_2])) - \mu(\tau_{k+1}([t_1,t_2]) \cap \theta([t_1,t_2])) \\ - \mu\Big((\tau_{k+1}([t_1,t_2]) \setminus \theta([t_1,t_2])) \cap \big(\bigcup_{j=1}^k \tau_j([t_1,t_2])\big)\Big) &> \frac{\varepsilon}{k+2}.\end{aligned}$$

By induction we have found therefore two sequences $\rho_n \to 0$ and $\tau_n \to \theta$ such that (5.15) holds for all $n \in \mathbb{N}$. This leads to a contradiction since the harmonic series diverges, while $\|\mu\| < \infty$. □

We can now find a stronger consequence of the definition (5.10), namely that different paths cannot cross in different directions as the following lemma ensures.

**Lemma 5.15.** *If $\eta \in \mathcal{M}^+_{\mu,1}(\Theta)$, then for no couple of paths $\theta, \tau \in \operatorname{spt}\eta$ one has (with suitable parameterizations) $\theta(t_1) = \tau(t_2)$ and $\theta(t_2) = \tau(t_1)$ for two time instants $0 \le t_1 < t_2 \le 1$.*

*Remark 5.16.* Notice that this assertion says, in words, that if two different paths in $\operatorname{spt}\eta$ pass through the same two points, they must do it in the same order. In particular, taking $\theta = \tau$, one gets that each $\theta \in \operatorname{spt}\eta$ is an arc (an injective curve), which also implies that $\ell(\theta) = \mathscr{H}^1(\theta)$; in other words, there are no paths in $\operatorname{spt}\eta$ with loops.

Note that, as in Lemma 4.17-iii), Lemma 4.18, Lemma 5.13 and Corollary 5.14, the most important part of this assertion is that we can establish the announced property for *any* path $\theta \in \operatorname{spt}\eta$, and not only for $\eta$-almost any.

*Proof (of Lemma 5.15).* Assume by contradiction that there are $\theta$ and $\tau$ as in the assertion being proved. Hence, $\theta([t_1,t_2])$ and $\tau([t_1,t_2])$ are contained in

$\Sigma$, since otherwise we would find a contradiction to the cyclical monotonicity property (4.43) with

$$x := \theta(0)\,, \qquad y := \theta(1)\,, \qquad x' := \tau(0)\,, \qquad y' := \tau(1)$$

and the optimal transport plan $\gamma = (p_0, p_1)_{\#}\eta$ associated to $\eta$.

By Corollary 5.14 there is a $\rho > 0$ such that for any

$$\tilde{\theta} \in \operatorname{spt} \eta \cap B_\Theta\big(\theta, \rho\big)$$

and

$$\tilde{\tau} \in \operatorname{spt} \eta \cap B_\Theta\big(\tau, \rho\big)$$

one has, minding that $\mathscr{H}^1 \geq \mu$ and that $\theta([t_1, t_2])$ and $\tau([t_1, t_2])$ are contained in $\Sigma$, the relationships

$$\begin{aligned} \mathscr{H}^1\big(\tilde{\theta}([t_1,t_2]) \cap \theta([t_1,t_2])\big) &\geq \mu\big(\tilde{\theta}([t_1,t_2]) \cap \theta([t_1,t_2])\big) > \frac{3}{4}\mu\big(\theta([t_1,t_2])\big) \\ &= \frac{3}{4}\mathscr{H}^1\big(\theta([t_1,t_2])\big)\,, \\ \mathscr{H}^1\big(\tilde{\tau}([t_1,t_2]) \cap \tau([t_1,t_2])\big) &\geq \mu\big(\tilde{\tau}([t_1,t_2]) \cap \tau([t_1,t_2])\big) > \frac{3}{4}\mu\big(\tau([t_1,t_2])\big) \\ &= \frac{3}{4}\mathscr{H}^1\big(\tau([t_1,t_2])\big)\,, \end{aligned} \tag{5.16}$$

with suitable parameterizations of $\tilde{\theta}$ and $\tilde{\tau}$. Let $t_1^{\tilde{\theta}}$ and $t_2^{\tilde{\theta}}$ (resp. $t_1^{\tilde{\tau}}$ and $t_2^{\tilde{\tau}}$) be the first and the last instants in $[t_1, t_2]$ when $\theta$ (resp. $\tau$) touches $\tilde{\theta}$ (resp. $\tilde{\tau}$). We denote then by $\hat{\theta}_1$ the shortest arc connecting $\theta(t_1)$ with $\theta(t_1^{\tilde{\theta}})$ inside $\theta([t_1, t_2])$, and by $\hat{\theta}_2$ the shortest arc connecting $\theta(t_2^{\tilde{\theta}})$ with $\theta(t_2)$ inside $\theta([t_1, t_2])$ (resp. by $\hat{\tau}_1$ the shortest arc connecting $\tau(t_1)$ with $\tau(t_1^{\tilde{\tau}})$ inside $\tau([t_1, t_2])$, and by $\hat{\tau}_2$ the shortest arc connecting $\tau(t_2^{\tilde{\tau}})$ with $\tau(t_2)$ inside $\tau([t_1, t_2])$): see Figure 5.2.

We now define in the obvious way $\alpha(\tilde{\theta}, \tilde{\tau})$ connecting $\tilde{\theta}(0)$ with $\tilde{\tau}(1)$ and $\beta(\tilde{\theta}, \tilde{\tau})$ connecting $\tilde{\tau}(0)$ with $\tilde{\theta}(1)$: formally,

$$\alpha(\tilde{\theta}, \tilde{\tau}) := \tilde{\theta}_{|[0,t_1]} \cdot \hat{\theta}_1 \cdot \hat{\tau}_2 \cdot \tilde{\tau}_{|[t_2,1]}\,, \qquad \beta(\tilde{\theta}, \tilde{\tau}) := \tilde{\tau}_{|[0,t_1]} \cdot \hat{\tau}_1 \cdot \hat{\theta}_2 \cdot \tilde{\theta}_{|[t_2,1]}\,.$$

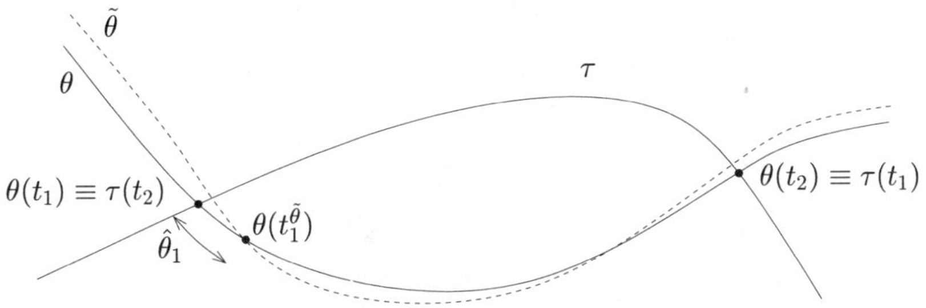


**Fig. 5.2** Situation in Lemma 5.15

In particular, the couple $(\alpha(\theta,\tau),\beta(\theta,\tau))$ is given by the formulae

$$\alpha = \theta_{|[0,t_1]} \cdot \tau_{|[t_2,1]}\,, \qquad \beta = \tau_{|[0,t_1]} \cdot \theta_{|[t_2,1]}\,.$$

The functions $\alpha, \beta$ are clearly Borel. Note also that

$$\delta_\mu(\tilde{\theta}) + \delta_\mu(\tilde{\tau}) = \delta_\mu(\alpha(\tilde{\theta},\tilde{\tau})) + \delta_\mu(\beta(\tilde{\theta},\tilde{\tau}))\,, \tag{5.17}$$

since $\theta([t_1,t_2])$ and $\tau([t_1,t_2])$ are contained in $\Sigma$. On the other hand,

$$\ell(\alpha(\tilde{\theta},\tilde{\tau})) + \ell(\beta(\tilde{\theta},\tilde{\tau})) < \ell(\tilde{\theta}) + \ell(\tilde{\tau})\,. \tag{5.18}$$

in view of (5.16). We consider now two measures $\eta_1$ and $\eta_2$ concentrated on sufficiently small balls centered in $\theta$ and $\tau$ respectively, so that $\alpha(\tilde{\theta},\tilde{\tau})$ and $\beta(\tilde{\theta},\tilde{\tau})$ are defined for any $\tilde{\theta} \in \operatorname{spt}\eta_1$, $\tilde{\tau} \in \operatorname{spt}\eta_2$, and with the property that

$$\|\eta_1\| = \|\eta_2\| =: \varepsilon > 0\,, \qquad \eta_1 + \eta_2 \leq \eta\,.$$

Thanks to the classical results on the isomorphism of measure spaces (see Theorem 2 in [53], or also Theorem 9 [61, Chapter 15]) we can take two Borel maps $\varphi_i : [0,\varepsilon] \to \Theta$, $i = 1, 2$ so that $\varphi_{i\#}\mathscr{L}_\varepsilon = \eta_i$, where $\mathscr{L}_\varepsilon$ stands for the Lebesgue measure on $[0,\varepsilon]$. Let us then consider the maps $\hat{\alpha}$, $\hat{\beta} : [0,\varepsilon] \to \Theta$ given by

$$\hat{\alpha}(t) := \alpha(\varphi_1(t),\varphi_2(t))\,, \qquad \hat{\beta}(t) := \beta(\varphi_1(t),\varphi_2(t))\,,$$

and finally define

$$\tilde{\eta} := \eta - \eta_1 - \eta_2 + \hat{\alpha}_\#\mathscr{L}_\varepsilon + \hat{\beta}_\#\mathscr{L}_\varepsilon\,.$$

By construction $\tilde{\eta}$ is an admissible t.p.m. and is contained in $\mathcal{M}^+_{\mu,0}(\Theta)$ by (5.17); moreover

$$\int_\Theta \ell(\theta)\,d\tilde{\eta}(\theta) < \int_\Theta \ell(\theta)\,d\eta(\theta)$$

due to (5.18), which gives a contradiction with the definition (5.10). □

Now, we see the importance of the maximization in (5.11), which provides a sort of "solidarity" between different paths in $\operatorname{spt}\eta$.

**Lemma 5.17.** *If $\eta \in \mathcal{M}^+_{\mu,2}(\Theta)$, given two points $x \neq y \in \Omega$ such that*

$$\eta\big(\{\sigma \in \Theta \,:\, x,\, y \in \sigma\}\big) > 0\,,$$

*then $\eta-$a.e. path passing through both $x$ and $y$ contains the same subpath $\sigma_{xy}$ between $x$ and $y$.*

*Proof.* Denote

$$\Theta_{xy} := \{\sigma \in \Theta : x, y \in \sigma\} \subseteq \Theta,$$

and for each $\sigma \in \Theta_{xy} \cap \operatorname{spt} \eta$ let $\tilde{\sigma}$ be the subpath of $\sigma$ connecting $x$ and $y$. Due to the assumption (5.7), for any $\sigma \in \Theta_{xy} \cap \operatorname{spt} \eta$ the subpath $\tilde{\sigma}$ is a geodesic between $x$ and $y$, so that $\delta_\mu(\tilde{\sigma})$ is a constant depending only on $x$ and $y$ (i.e. independent on $\sigma$). We obtain the thesis showing that, by the maximization in (5.11), one of these geodesics is chosen by $\eta$-a.e. path in $\Theta_{xy}$. To this aim we first show that there is a constant $c$ such that $\mathscr{H}^1(\tilde{\sigma}) = c$ for $\eta$-a.e. $\sigma \in \Theta_{xy}$. In fact, otherwise there would exist two constants $c_1 < c_2$ such that

$$\eta(\{\sigma \in \Theta_{xy} : \mathscr{H}^1(\tilde{\sigma}) = c_i\}) > 0, \qquad i = 1, 2.$$

Define the function $f\colon \Theta \to \Theta$ according to the relationship

$$f(\sigma) := \begin{cases} \sigma_1, \ \sigma \in \Theta_{xy} \cap \operatorname{spt} \eta, \mathscr{H}^1(\tilde{\sigma}) = c_2, \\ \sigma, \ \text{otherwise}, \end{cases} \tag{5.19}$$

where $\sigma_1 \in \Theta_{xy} \cap \operatorname{spt} \eta$ is some chosen path with $\mathscr{H}^1(\tilde{\sigma}) = c_1$. We have then that $\eta' := f_{\#}\eta$ is still an optimal t.p.m., but clearly

$$\int_\Theta \mathscr{H}^1(\theta)\, d\eta' < \int_\Theta \mathscr{H}^1(\theta)\, d\eta,$$

contradicting the assumption $\eta \in \mathcal{M}^+_{\mu,1}(\Theta)$ (recall that $\mathscr{H}^1(\theta) = \ell(\theta)$ for all $\theta \in \operatorname{spt} \eta$ according to Lemma 5.15).

Now, since both $\delta_\mu(\tilde{\sigma})$ and $\mathscr{H}^1(\tilde{\sigma})$ are $\eta$-a.e. constant for $\sigma \in \Theta_{xy}$, we get that so is

$$\mu(\tilde{\sigma}) = \mathscr{H}^1(\tilde{\sigma}) - \delta_\mu(\tilde{\sigma}).$$

Choose now an arbitrary

$$\sigma' \in \Theta_{xy} \cap \operatorname{spt} \eta$$

and let $\hat{\eta}$ be the t.p.m. that leaves all the paths of $\operatorname{spt} \eta$ unchanged, except those connecting $x$ and $y$, the latter being forced to follow $\tilde{\sigma}$ between $x$ and $y$. Formally, we define the function $g\colon \Theta \to \Theta$ by setting $g(\sigma)$ for $\sigma \in \Theta_{xy} \cap \operatorname{spt} \eta$ to be the composition of the arc of $\sigma$ between $\sigma(0)$ and $x$, $\tilde{\sigma}'$, and the arc of $\sigma$ between $y$ and $\sigma(0)$ (for all the other paths $\sigma$ we set $g(\sigma) := \sigma$). A straightforward computation ensures that

$$\hat{\eta} := g_{\#}\eta \in \mathcal{M}^+_{\mu,1}(\Theta),$$

while since $\alpha_{\hat{\eta}}$ is just a redistribution of $\alpha_\eta$, one has using the strict convexity of $\alpha_\eta(x)^2$ that

$$\int_\Sigma \alpha_{\hat{\eta}}(z)^2\, d\mathscr{H}^1(z) < \int_\Sigma \alpha_\eta(z)^2\, d\mathscr{H}^1(z),$$

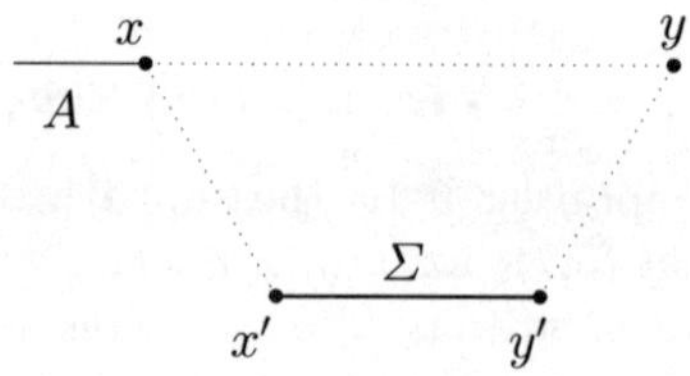


**Fig. 5.3** Situation in Example 5.19

unless $\eta$-a.e. path connecting $x$ and $y$ already follows a unique path, which concludes the proof. □

*Remark 5.18.* Notice that the path $\sigma_{xy}$ is well-defined only for such pairs $(x, y)$ that the set $\Theta_{xy}$ in the above proof is not $\eta$-negligible. Moreover, if $\sigma_{xy}$ is defined and $z$, $w \in \sigma_{xy}$, then $\sigma_{zw}$ is defined as well and it is the subpath of $\sigma_{xy}$ connecting $z$ and $w$.

As the Example 5.19 below shows, the above result is not true for *all* the paths in $\operatorname{spt}\eta$ connecting $x$ and $y$, but only for $\eta$-a.e.. This is in contrast with all the other properties enlisted before, which were true for all the paths in $\operatorname{spt}\eta$, and not just for $\eta$-almost all.

*Example 5.19.* Consider the following situation, shown in Figure 5.3: let

$$f^+ = \mathscr{H}^1 \llcorner A + \delta_x + M\delta_{x'} \qquad\qquad f^- = 2\delta_y + M\delta_{y'} ;$$

here $A$ is a segment of unit length, having $x$ as a right endpoint. The geometry is chosen so that

$$|x - y| = |x - x'| + |y - y'| .$$

If $M$ is large enough, it is easy to define $H$ in such a way that the unique optimal measure is $\mu := \mathscr{H}^1 \llcorner \Sigma$, with $\Sigma := [x', y']$ being the line segment connecting $x'$ to $y'$, and that there is a unique optimal t.p.m. $\eta \in \mathcal{M}^+_{\mu,1}(\Theta)$ (hence $\eta \in \mathcal{M}^+_{\mu,2}(\Theta)$) which moves the mass as follows:

- $M\delta_{x'}$ goes on $M\delta_{y'}$ along $\Sigma$;
- $\delta_x$ goes on $\delta_y$ via the line segment $[x, y]$: using $\Sigma$ would have the same cost but would contradict the fact that $\eta \in \mathcal{M}^+_{\mu,1}(\Theta)$, namely the minimization (5.10);
- $\mathscr{H}^1 \llcorner A$ goes on $\delta_y$ moving each point $z \in A$ along the line segment $[z, x']$, then following the network between $x'$ and $y'$, and finally along the line segment $[y', y]$.

By construction, the path $\theta$ connecting $x$ to $y$ and passing through $\Sigma$ belongs to the support of $\eta$, since $\operatorname{spt}\eta$ is closed and $x \in A$. On the other hand $\sigma_{xy}$ is the line segment $[x, y]$. This shows that the assertion of Lemma 5.17 cannot be strengthened replacing "$\eta$-a.e. path connecting $x$ and $y$" with "each path in $\operatorname{spt}\eta$ connecting $x$ and $y$". However, notice that $\sigma_{x'y'}$ coincides with the restriction of $\theta$ between $x'$ and $y'$, according to Lemma 5.20 below.

Nevertheless, we can show that a sort of *everywhere solidarity*, i.e. that *all* the paths in $\operatorname{spt}\eta$ (instead of $\eta-$almost all) follow the path $\sigma_{xy}$ between $x$ and $y$, is true where there is some railway. In the following, when a path $\theta \in \Theta$ and two points $x, y \in \theta$ are given, we will write $\widetilde{xy}$ to denote the subpath of $\theta$ connecting $x$ to $y$.

**Lemma 5.20.** *Let $\eta \in \mathcal{M}^+_{\mu,2}(\Theta)$, let $\theta \in \operatorname{spt}\eta$ be a path such that $\mu(\theta) > 0$, and let $x'$ and $y'$ be such points that $\widetilde{x'y'}$ is the shortest subpath of $\theta$ satisfying $\mu(\widetilde{x'y'}) = \mu(\theta)$. Then, for any two points $x$, $y$ belonging to the open subpath $\widetilde{x'y'}$, the path $\sigma_{xy}$ introduced in Lemma 5.17 is well-defined and coincides with $\widetilde{xy}$.*

*Proof.* Let us define

$$B := \left\{ z \in \widetilde{x'y'} : \liminf_{n\to\infty} \frac{\eta\big(\{\sigma \in B_\Theta(\theta, 1/n) : z \in \sigma\}\big)}{\eta\big(B_\Theta(\theta, 1/n)\big)} \le \frac{1}{2} \right\}$$

the set of "bad" points of $\widetilde{x'y'}$, i.e. those points in $\widetilde{x'y'}$ such that it happens that only few paths close to $\theta$ pass through them. Thanks to Corollary 5.14 we have $\mu(B) = 0$. Indeed, if $\mu(B) > 0$, then by definition there is a sequence $\rho_\nu \to 0$ and a constant $C > 0$ such that

$$\mu\left(\left\{ z \in \widetilde{x'y'} : \frac{\eta\big(\{\sigma \in B_\Theta(\theta, \rho_\nu) : z \in \sigma\}\big)}{\eta\big(B_\Theta(\theta, \rho_\nu)\big)} > \frac{1}{2} \right\}\right) \le \mu(\theta) - C$$

for all $\nu \in \mathbb{N}$, which contradicts Corollary 5.14. Therefore, for $\mathscr{H}^1-$a.e. point $z \in \widetilde{x'y'} \cap \Sigma$, more than 50% (with respect to $\eta$) of the paths which are close to $\theta$ pass through $z$. Take then two such points $x$ and $y$: for any $\rho > 0$ the measure $\eta$ of the paths in $B_\Theta(\theta, \rho)$ passing through both $x$ and $y$ is strictly positive. Then $\sigma_{xy}$ is well-defined. We prove now $\sigma_{xy} = \widetilde{xy}$. Indeed, if it were not so, then one would have

$$d_\Theta\big(\widetilde{xy}, \sigma_{xy}\big) =: \rho > 0$$

which leads to a contradiction, since by Lemma 5.17, $\eta$-a.e. path passing through both $x$ and $y$ must follow $\sigma_{xy}$, which is impossible for the paths belonging to $B_\Theta\big(\theta, \rho/2\big)$, that are not $\eta$-negligible. Finally, since the points $x$ and $y$ can be chosen arbitrarily close to $x'$ and $y'$ by the definition of $x'$ and $y'$ and by the fact that $\mu(B) = 0$, the thesis follows by Remark 5.18. □

## 5.3 Closedness of Optimal Sets

In this section we derive the closedness of an optimal set $\Sigma$ by the concavity of $H$. More precisely, we show that if $H$ is concave (resp. strictly concave) then there exists a closed optimal set $\Sigma$ (resp. *any* optimal measure corresponds to an essentially closed set). Moreover, we present an example showing that the concavity of $H$ is crucial for such results. Recall that we are assuming that $A(t) = t$ and $B(t) = 0$ for all $t \in \mathbb{R}^+$.

**Theorem 5.21.** *If $H$ is concave, then there exists a closed optimal set $\Sigma$. Moreover, if $H$ is strictly concave then any optimal measure $\mu$ corresponds to an essentially closed optimal set (in particular, every optimal set is essentially closed).*

*Proof.* Take an optimal measure $\mu$ and a t.p.m. $\eta$ minimizing $C_\mu$, and define, in accordance with Theorem 5.5, $r > 0$ to be the maximal constant satisfying (5.2). By the maximality of $r$, for any $\varepsilon > 0$ the set

$$\Sigma_\varepsilon := \big\{\alpha_\eta(x) \geq r + \varepsilon\big\}$$

is such that $\mu - \mathscr{H}^1 \llcorner \Sigma_\varepsilon > 0$. Defining then $\mu_\varepsilon := \mathscr{H}^1 \llcorner \Sigma_\varepsilon$ and recalling (5.8), we immediately deduce

$$C_{\mu_\varepsilon}(\eta) \leq C_\mu(\eta) + \|\mu - \mu_\varepsilon\| \big(r + \varepsilon\big) . \tag{5.20}$$

On the other hand, since $H$ is concave we can evaluate

$$H\big(\mathscr{H}^1(\Sigma_\varepsilon)\big) = H\big(\|\mu\| - (\|\mu - \mu_\varepsilon\|)\big) \leq H(\|\mu\|) - H'_-(\|\mu\|)\|\mu - \mu_\varepsilon\| , \tag{5.21}$$

where $H'_-(s)$ is the left derivative of $H$ at $s \in \mathbb{R}^+$, which is well-defined since $H$ is concave. Since $MK(\mu) = C_\mu(\eta)$ because $\eta$ is an optimal t.p.m., $\mathfrak{F}(\mu) \leq \mathfrak{F}(\mu_\varepsilon)$ because $\mu$ is an optimal measure, and $MK(\mu_\varepsilon) \leq C_{\mu_\varepsilon}(\eta)$, we deduce by (5.20) and (5.21) that

$$r + \varepsilon \geq H'_-\big(\|\mu\|\big) .$$

Since $\varepsilon$ is arbitrary, we obtain

$$H'_-\big(\|\mu\|\big) \leq r . \tag{5.22}$$

Let us define now

$$\overline{\Sigma} := \big\{\alpha_\eta(x) \geq r\big\} ,$$

which is a closed set by the upper semicontinuity of $\alpha_\eta$; notice that $\overline{\Sigma}$ contains the support of $\mu$, but it may also be strictly bigger. Let also $\bar\mu := \mathscr{H}^1 \llcorner \overline{\Sigma}$; if $\|\bar\mu - \mu\| = 0$, then $\overline{\Sigma}$ is an optimal closed set and hence there is nothing to prove. Otherwise, again by (5.8) we know that

$$C_{\bar{\mu}}(\eta) = C_\mu(\eta) - \int_{\overline{\Sigma}} \alpha_\eta(x)\, d(\bar{\mu} - \mu)(x)\,.$$

By definition, $\alpha_\eta \geq r$ on $\operatorname{spt}(\bar{\mu} - \mu)$, so that the last estimate becomes

$$C_{\bar{\mu}}(\eta) \leq C_\mu(\eta) - r\,\|\bar{\mu} - \mu\|\,.$$

On the other hand, again by the concavity of $H$ one has

$$H\big(\mathscr{H}^1(\overline{\Sigma})\big) = H\big(\|\bar{\mu}\|\big) \leq H\big(\|\mu\|\big) + H'_-\big(\|\mu\|\big)\,\|\bar{\mu} - \mu\|\,. \tag{5.23}$$

Summarizing, we obtain

$$\begin{aligned} \mathfrak{F}(\bar{\mu}) &\leq C_{\bar{\mu}}(\eta) + H\big(\mathscr{H}^1(\overline{\Sigma})\big) \\ &\leq C_\mu(\eta) + H(\|\mu\|) + \big(H'_-(\|\mu\|) - r\big)\,\|\bar{\mu} - \mu\| \\ &= \mathfrak{F}(\mu) + \big(H'_-(\|\mu\|) - r\big)\,\|\bar{\mu} - \mu\| \leq \mathfrak{F}(\mu)\,, \end{aligned} \tag{5.24}$$

the last inequality being valid in view of (5.22). Since $\mu$ is optimal, we have proved that also $\bar{\mu} = \mathscr{H}^1 \llcorner \overline{\Sigma}$ is an optimal measure and then that $\overline{\Sigma}$ is a closed optimal set: thus the existence of some closed optimal set is established.

Concerning the second part of the claim, if $H$ is strictly concave and $\|\bar{\mu}-\mu\| > 0$ then the first inequality in (5.23) becomes strict; as a consequence, also the inequality (5.24) is strict and this gives a contradiction with the optimality of $\mu$. Hence, we may conclude that $\mu = \bar{\mu}$ and therefore a generic optimal measure corresponds to a closed optimal set. □

Recall that Theorem 4.26 already ensures that all the optimal measures are represented by sets if $D(\cdot, b)$ is strictly concave. The above Theorem 5.21 says that in the case $A(t) = t$, $B(t) = 0$, in which $D(a, b) = a$ is not strictly concave, it is still true that all the optimal measures are sets whenever $H$ is strictly concave.

The following example shows that the hypothesis of concavity of $H$ in Theorem 5.21 is essential.

*Example 5.22.* Let

$$\Omega := [-2, 2] \times [-1, 1] \subseteq \mathbb{R}^2\,,$$

and let $a_n < b_n$ be such that all the open segments $(a_n, b_n)$ are disjoint, and their union $D := \cup_j (a_j, b_j)$ contains the set of all the rational numbers in $[-1, 1]$ and has length 1. Define now

$$f^+ := \sum_{n \in \mathbb{N}} 2^{-n} \delta_{(a_n, 0)}\,, \qquad\qquad f^- := \sum_{n \in \mathbb{N}} 2^{-n} \delta_{(b_n, 0)}\,,$$

let $\theta_n$ be the line segment connecting $(a_n, 0)$ to $(b_n, 0)$, and let $\eta$ be any t.p.m. related to $f^+$ and $f^-$. Since for any $n$ one has

$$f^+([-2,a_n]) = \sum_{\{j: a_j \le a_n\}} 2^{-j} = \sum_{\{j: a_j < b_n\}} 2^{-j} > \sum_{\{j: b_j < a_n\}} 2^{-j}$$
$$= f^-([-2,a_n]),$$

there must be some $\theta \in \operatorname{spt}\eta$ connecting some $a_i \le a_n$ to some $b_j \ge b_n$. This implies that the projection on the first axis of the set

$$S_\eta = \bigcup_{\theta \in \operatorname{spt}\eta} \theta \subseteq \mathbb{R}^2$$

contains the whole segment $[a_n, b_n]$. Let now

$$\bar\eta := \sum_{n\in\mathbb{N}} 2^{-n}\,\delta_{\theta_n},$$

which is clearly a t.p.m. related to $f^+$ and $f^-$, let $\Sigma := \cup_{n\in\mathbb{N}}\theta_n$ and define the function $H$ by the relationship

$$H(l) := \begin{cases} 0,\, l \le 1, \\ \infty,\, l > 1. \end{cases}$$

Obviously, $H$ is not concave. Notice that $C_\Sigma(\bar\eta) = 0$ and that $H(\mathscr{H}^1(\Sigma)) = 0$, so that $\mathfrak{F}(\Sigma) = 0$. Consider now any $\mu \le \mathscr{H}^1$ such that $\mathfrak{F}(\mu) = 0$. One has then $\|\mu\| \le 1$ and $C_\mu(\eta) = 0$ for some t.p.m. $\eta$. This implies $\mu(\theta) \ge \mathscr{H}^1(\theta)$ for $\eta$-a.e. $\theta \in \Theta$, and hence, minding the semicontinuity of $\mu$ and $\mathscr{H}^1$ over $\Theta$, actually for all $\theta \in \operatorname{spt}\eta$. Therefore, one has

$$\mu \llcorner \theta = \mathscr{H}^1 \llcorner \theta, \qquad\qquad \text{for all } \theta \in \operatorname{spt}\eta$$

(because $\mu \le \mathscr{H}^1$). Since for any $n$ there is a path $\theta_n \in \operatorname{spt}\eta$ connecting some $a_i \le a_n$ with some $b_j \ge b_n$, then calling $\pi$ the projection on the first axis one has

$$1 \ge \|\mu\| \ge \mu(\cup_n\theta_n) = \mathscr{H}^1(\cup_n\theta_n) \ge \mathscr{H}^1\big(\pi(\cup_n\theta_n)\big) \ge \mathscr{H}^1\big(\cup_n\,[a_n,b_n]\big)$$
$$= \sum_n |b_n - a_n| = 1,$$

from which we derive that $\mu = \mathscr{H}^1 \llcorner \Sigma$. Summarizing, $\mathfrak{F}(\mu) > 0$ for any $\mu \ne \mathscr{H}^1 \llcorner \Sigma$. This shows that $\mathscr{H}^1 \llcorner \Sigma$ is the unique optimal measure, so that $\Sigma$ is the unique (up to a an $\mathscr{H}^1$-negligible set) optimal set; however, this set is not essentially closed: indeed, it is an open set of length 1 dense in $[-1,1]$.

## 5.4 Number of Connected Components of Optimal Sets

As already pointed out, it would be interesting to understand if there is an optimal set $\Sigma$ having finitely many, or at least countably many, connected components. We however can immediately notice that this is not always the case, according to the following counterexample similar to Example 5.22.

*Example 5.23.* Let $\Omega = [-2,2] \times [-1,1] \subseteq \mathbb{R}^2$. Consider now the following construction of a Cantor-type set in $[0,1]$ having strictly positive length. Given a sequence $\{p_n\}_{n\in\mathbb{N}}$ with $0 < p_n < 1/2$ for any $n$, we first divide $[0,1]$ in the three segments

$$[\alpha_{1,1}, \beta_{1,1}], \qquad [\beta_{1,1}, \alpha_{1,2}], \qquad [\alpha_{1,2}, \beta_{1,2}]$$

having length $p_1$, $1-2p_1$ and $p_1$ respectively, defining

$$\alpha_{1,1} := 0, \qquad \beta_{1,1} := p_1, \qquad \alpha_{1,2} := 1 - p_1, \qquad \beta_{1,2} := 1.$$

We continue by induction, defining $\alpha_{n,m}$ and $\beta_{n,m}$ for any $n \in \mathbb{N}$ and any $1 \leq m \leq 2^n$ in such a way that any segment $[\alpha_{n,m}, \beta_{n,m}]$ is divided in the three segments

$$[\alpha_{n+1,2m-1}, \beta_{n+1,2m-1}], \qquad [\beta_{n+1,2m-1}, \alpha_{n+1,2m}], \qquad [\alpha_{n+1,2m}, \beta_{n+1,2m}],$$

the lengths of which are in the ratio $p_{n+1}$, $1-2p_{n+1}$ and $p_{n+1}$ respectively. We define then $K_n$ as the closed set given by the union of the $2^n$ closed segments $[\alpha_{n,m}, \beta_{n,m}]$ and $K$ as the intersection of the sets $K_n$. Notice that the classical Cantor set corresponds to the choice $p_n = 1/3$ for any $n$. Similarly to the Cantor set, $K$ is a closed set containing more than countably many points and is totally disconnected, that is, all the connected components of $K$ reduce to a single point. Moreover,

$$\mathscr{H}^1(K) = \lim_{n\to\infty} \mathscr{H}^1(K_n)$$

and

$$\mathscr{H}^1(K_{n+1}) = 2p_{n+1}\mathscr{H}^1(K_n),$$

so that a suitable choice of the coefficients $p_n$ allows $\mathscr{H}^1(K)$ to be any number in $[0,1)$. We consider then a sequence $\{p_n\}$ such that $\mathscr{H}^1(K) = 1/2$. Now, we define $f^+$ and $f^-$ as follows:

$$f^+ := \sum_{n\in\mathbb{N}} \sum_{1\leq m\leq 2^n} 4^{-n}\delta_{(\alpha_{n,m},0)}, \qquad f^- := \sum_{n\in\mathbb{N}} \sum_{1\leq m\leq 2^n} 4^{-n}\delta_{(\beta_{n,m},0)}.$$

It is easily noticed that $f^+$ and $f^-$ are two probability measures and that, setting $\theta_{n,m}$ to be the segment $[\alpha_{n,m}, \beta_{n,m}]$, the measure

$$\eta := \sum_{n\in\mathbb{N}} \sum_{1\le m\le 2^n} 4^{-n}\delta_{\theta_{n,m}}$$

is a t.p.m. related to $f^+$ and $f^-$. We set now

$$H(s) := \begin{cases} 0 & \text{for } 0 \le s \le 1/2\,, \\ \infty & \text{for } s > 1/2\,. \end{cases}$$

It is immediate to notice that any optimal measure $\mu$ must be supported in $[0,1] \times \{0\}$. Moreover, arguing as in Example 5.22, one has that given any measure $\mu$ supported in $[0,1]\times\{0\}$, all the paths $\theta_{n,m} \in \operatorname{spt}\eta'$ for every t.p.m. $\eta'$, and hence, the t.p.m. $\eta$ defined above is optimal for the cost $C_\mu$. It is also clear that the function $\alpha_\eta$ achieves its maximum exactly in the set $K$. Hence, the measure $\mu := \mathscr{H}^1 \llcorner K$ is the unique optimal measure, by Theorem 5.5 and by the choice of $H$. Therefore, $K$ is an optimal set which has more than countably many connected components, since it is totally disconnected and uncountable. The optimal sets are exactly those coinciding with $K$ up to $\mathscr{H}^1$–negligible sets; it is easily noticed that any such set $K'$ is still totally disconnected.

*Remark 5.24.* By a more careful construction, replacing Dirac masses in the above example by smooth functions with "small" supports, it is possible to obtain a counterexample as before with absolutely continuous measures $f^\pm$ with densities belonging even to $C^\infty$.

We give now conditions for the existence of an optimal set having finite or countably many connected components: as in Theorem 5.21, a concavity hypothesis on $H$ will be essential. First, we prove the following *a posteriori* result.

**Lemma 5.25.** *Let $\mu$ be an optimal measure, $\eta$ an ordered optimal t.p.m. for $\mu$, and assume that the constant $r$ satisfying (5.2) is not unique. Then,*

$$\mu = \mathscr{H}^1 \llcorner \Sigma$$

*for some closed set $\Sigma$ such that $\Sigma\cap\theta$ has finitely many connected components for all $\theta \in \operatorname{spt}\eta$.*

*Proof.* First of all we stress that, since the constant $r$ in (5.2) is not unique, then $\mu$ corresponds to a set: indeed, (5.2) is satisfied by all $r_1 < r < r_2$, so that

$$\mathscr{H}^1\big(\{\alpha_\eta(x) = r\}\big) = 0$$

for any such $r$ and thus $\mu = \mathscr{H}^1 \llcorner \Sigma$ with $\Sigma = \big\{\alpha_\eta(x) \ge r\big\}$. Notice that $\Sigma$ is closed since, by Lemma 5.4, $\alpha_\eta$ is u.s.c..

Take now a single path $\theta \in \operatorname{spt}\eta$: we define $\sigma$ as the shortest subpath of $\theta$ such that

$$\sigma \cap \Sigma = \theta \cap \Sigma\,.$$

We claim first that the function

$$\alpha_\eta \circ \sigma : t \in [0,1] \mapsto \alpha_\eta(\sigma(t))$$

has bounded variation. This amounts to proving that the weak derivative $(\alpha_\eta \circ \sigma)'$ is a finite measure, where the derivative $(\alpha_\eta \circ \sigma)'$ is intended as the weak limit of

$$\frac{\alpha_\eta(\sigma(s+\varepsilon)) - \alpha_\eta(\sigma(s))}{\varepsilon}$$

as $\varepsilon \to 0$. Define for this purpose $\Theta'$ to be the set of all $\alpha \in \operatorname{spt}\eta$ such that $\mu(\alpha \cap \sigma) > 0$. By Lemma 5.20 we deduce that, for $\eta$-a.e. $\alpha \in \Theta'$, defining $p(\alpha)$ and $q(\alpha)$ to be the first and the last point of $\alpha$ intersecting $\sigma$ respectively, the subpath $\widetilde{pq}$ of $\alpha$ connecting $p(\alpha)$ and $q(\alpha)$ is entirely contained in $\sigma$. In other words, there are almost no paths entering in $\sigma$, then exiting, then entering again, according to what we called "solidarity" in Section 5.2. As an easy consequence, for any $x$ and $y$ in $\sigma$,

$$\alpha_\eta(y) - \alpha_\eta(x) = \eta\big(\{\alpha \in \Theta' : \ p(\alpha) \in \widetilde{xy}\}\big) - \eta\big(\{\alpha \in \Theta' : \ q(\alpha) \in \widetilde{xy}\}\big),$$

that is,

$$\alpha'_\eta = p_\# \eta - q_\# \eta .$$

Hence, $(\alpha_\eta \circ \sigma)'$ is a finite (signed) measure and the claim is shown.

Now, take $r_1 > r_2$ such that (5.2) is satisfied both for $r_1$ and for $r_2$, and take a point $x \in \partial\big(\Sigma \cap \sigma\big)$ where $\partial$ stands for the boundary relative to $\sigma$. Since $x$ belongs to the relative boundary of $\Sigma$ in $\sigma$, then for any $\varepsilon > 0$ there is a point $z_\varepsilon \in \sigma \cap \Sigma$ and a point $w_\varepsilon \in \sigma \setminus \Sigma$ having distance less than $\varepsilon$ from $x$. Using (5.2) with the points $z_\varepsilon$ and $w_\varepsilon$, and the constants $r_1$ and $r_2$ respectively, we notice that $\alpha_\eta(z_\varepsilon) \geq r_1$ and that $\alpha_\eta(w_\varepsilon) \leq r_2$. Since $\varepsilon$ is arbitrary, it follows that $\alpha_\eta$ has, at the point $x$, a jump of at least $r_1 - r_2$, i.e.,

$$\big|(\alpha_\eta \circ \sigma)'(\{x\})\big| \geq r_1 - r_2 .$$

Since, as proved above, $\alpha_\eta \circ \sigma$ is a function of bounded variation, we have that $\partial\Sigma$ contains only finitely many points. Therefore, $\Sigma \cap \sigma$ has only a finite number of connected components and the proof is concluded. □

**Corollary 5.26.** *If $\mu$ is an optimal measure, $\eta$ is an ordered optimal t.p.m. for $\mu$, and the constant $r$ satisfying (5.2) is not unique, then $\mu = \mathscr{H}^1 \llcorner \Sigma$ for some closed set $\Sigma$ having countably many connected components.*

*Proof.* By the non-uniqueness assumption, (5.2) holds with some $r > 0$. Hence, by Lemma 5.7, $\mu$ is concentrated in the union of at most countably many geodesics, which are Lipschitz curves of uniformly bounded length. Therefore, $\Sigma$ is contained in the (at most countable) union of $\theta_i$, where each $\theta_i$ is contained in $\operatorname{spt}\eta$ and $\mathscr{H}^1(\Sigma \cap \theta_i) > 0$ for any $i$. The conclusion now follows directly from Lemma 5.25. □

*Remark 5.27.* Arguing as in Corollary 5.26, we can deduce that $\Sigma$ has a finite number of connected components, instead of a countable number, whenever we know that $\Sigma$ is contained in finitely many geodesics with respect to $d_\Sigma$. For instance, this is the case when the conditions of Theorem 6.1 hold: in this case $\Sigma$ is contained in finitely many geodesics, and hence the non-uniqueness of $r$ allows us to obtain that $\Sigma$ has finitely many connected components.

We prove now a result ensuring the existence of at most countably many connected components of an optimal set $\Sigma$ whenever $H$ is concave.

**Theorem 5.28.** *If $H$ is concave, then there exists an optimal set $\Sigma$ having at most countably many connected components. Moreover, if $H$ is strictly concave, then any optimal measure $\mu$ corresponds to an optimal set having at most countably many connected components.*

*Proof.* Consider an optimal measure $\mu$, an ordered optimal t.p.m. $\eta$ for $C_\mu$ and a constant $r$ satisfying (5.2). We show the first part of the claim considering separately three possible cases.

Case I. $H'_+(\|\mu\|) < r$.
Let us write

$$H'_+(\|\mu\|) = r - \varepsilon$$

with some $\varepsilon > 0$: we claim that the measure

$$\tilde{\mu} := \mathscr{H}^1 \llcorner \{x \in \Omega : r - \varepsilon < \alpha_\eta(x)\}$$

coincides with $\mu$. Indeed, clearly $\tilde{\mu} \geq \mu$. Moreover, if

$$\|\tilde{\mu} - \mu\| = \delta > 0,$$

then by the concavity of $H$ one has

$$H(\|\tilde{\mu}\|) - H(\|\mu\|) \leq (r - \varepsilon)\delta,$$

while on the other hand, by (5.8) we deduce

$$C_{\tilde{\mu}}(\eta) < C_\mu(\eta) - (r - \varepsilon)\delta.$$

Summarizing, we have $\mathfrak{F}(\tilde{\mu}) < \mathfrak{F}(\mu)$ which contradicts the optimality of $\mu$. Therefore, we obtain that $\tilde{\mu} = \mu$, but then any constant between $r - \varepsilon$ and $r$ suits for (5.2). Thus Corollary 5.26 ensures that $\mu$ corresponds to a set with countably many connected components.

Case II. $H'_-(\|\mu\|) > r$.
This case is analogous to the previous one: we write

$$H'_-(\|\mu\|) = r + \varepsilon$$

for some $\varepsilon > 0$ and set

$$\tilde{\mu} := \mathscr{H}^1 \llcorner \{x \in \Sigma : r + \varepsilon < \alpha_\eta(x)\} .$$

This time $\tilde{\mu} \leq \mu$, and exactly as before we find that $\tilde{\mu} = \mu$ since otherwise $\tilde{\mu}$ would be strictly better than the optimal measure $\mu$. Again, the constant suitable for (5.2) is not unique and Corollary 5.26 allows to conclude.

Case III. $H'(\|\mu\|) = r$.
In this last case, we can assume that $\mu = \mathscr{H}^1 \llcorner \Sigma$ for some Borel set $\Sigma$ thanks to Theorem 4.26. Noticing that $r > 0$ since by hypothesis $H$ is concave and unbounded, we may argue as in Corollary 5.26 and deduce that

$$\Sigma \subseteq \bigcup_i \theta_i$$

for countably many geodesics $\theta_i \in \operatorname{spt} \eta$. Denote by $\sigma_i$ the smallest subpath of $\theta_i$ such that

$$\sigma_i \cap \Sigma = \theta_i \cap \Sigma ,$$

and write $\Sigma_i^\circ$ to denote the interior part of $\Sigma \cap \sigma_i$ relative to $\sigma_i$. Define then the "boundary" $\partial\Sigma$ of $\Sigma$ to be the union of the boundaries $\Sigma \cap \sigma_i$ relative to $\sigma_i$. Take then a point $x \in \partial\Sigma$ with $\alpha_\eta(x) \neq r$: arguing as at the end of the proof of Lemma 5.25, we deduce that it is a discontinuity point for $\alpha_\eta$. But since $\alpha_\eta$ is a function of bounded variation on each $\sigma_i$, we know that it can have at most countably many discontinuity points, and hence we deduce that $\alpha_\eta(x) = r$ for $\mathscr{H}^1$–a.e. point of $\partial\Sigma$. Considering then $\Sigma^\circ := \Sigma \setminus \partial\Sigma$ and writing $\delta := \mathscr{H}^1(\Sigma \cap \partial\Sigma)$, one has

$$H(\mathscr{H}^1(\Sigma^\circ)) \leq H(\mathscr{H}^1(\Sigma)) - r\delta$$

again by the concavity of $H$. Moreover

$$C_{\Sigma^\circ}(\eta) = C_\Sigma(\eta) + r\delta ,$$

and therefore we find $\mathfrak{F}(\Sigma^\circ) \leq \mathfrak{F}(\Sigma)$. Hence, $\Sigma^\circ$ is an optimal set. But for each $i$ the set $\Sigma_i^\circ$ is open relative to $\theta_i$ and, since $\Sigma^\circ = \bigcup_i \Sigma_i^\circ$, then $\Sigma^\circ$ is a countable union of open subpaths of the curves $\theta_i$. Therefore $\Sigma^\circ$ is an optimal set having at most countably many connected components.

We have then shown the first part of the Theorem. Consider now what happens if $H$ is strictly concave: we need to prove that in this case the generic optimal measure $\mu$ corresponds to a set, and that this set has at most countably many connected components. If one of the two cases I or II above occurs, we have already shown that $\mu$ corresponds to a set with countably many connected components without any need of the strict concavity of $H$, hence there is nothing to prove. We need therefore to consider what happens in the case III. We affirm what follows:

*If $H$ is strictly concave and the case III above occurs, then*

$$\mathscr{H}^1\left(\{\alpha_\eta(x) = r\}\right) = 0\,.$$

Notice that this claim will give the thesis: indeed, if

$$\mathscr{H}^1(\{\alpha_\eta(x) = r\}) = 0$$

then

$$\{\alpha_\eta(x) \geq r\} = \{\alpha_\eta(x) > r\}$$

in the $\mathscr{H}^1$–sense, so by (5.2) we deduce that $\mu = \mathscr{H}^1 \llcorner \Sigma$ with

$$\Sigma := \{\alpha_\eta(x) \geq r\}\,.$$

In particular, since we proved above that $\alpha_\eta(x) = r$ for all but countably many $x \in \partial\Sigma$, one has

$$\{\alpha_\eta(x) > r\} \subseteq \Sigma^\circ \subseteq \{\alpha_\eta(x) \geq r\}\,,$$

where the first inclusion is true up to a countable (hence $\mathscr{H}^1$-negligible) set. Thus, one has

$$\Sigma = \{\alpha_\eta(x) \geq r\} = \{\alpha_\eta(x) > r\} = \Sigma^\circ$$

where the equalities are understood up to an $\mathscr{H}^1$-negligible set, and hence $\mu = \mathscr{H}^1 \llcorner \Sigma^\circ$ which gives the thesis.

We are then reduced to prove the above claim. For this purpose, we assume

$$\mathscr{H}^1(\{\alpha_\eta(x) = r\}) > 0$$

and we find a contradiction in the following two cases, at least one of which has to occur.

Case III.A. $\|\mathscr{H}^1 \llcorner \{\alpha_\eta(x) = r\} \wedge \mu\| = \delta > 0\,.$
In this case, define

$$\tilde{\mu} := \mu - \left(\mathscr{H}^1 \llcorner \{\alpha_\eta(x) = r\} \wedge \mu\right).$$

One has that

$$C_{\tilde{\mu}}(\eta) = C_\mu(\eta) + r\delta, \text{ and } H(\|\tilde{\mu}\|) < H(\|\mu\|) - r\delta\,,$$

because $H$ is strictly concave and $\delta > 0$. This implies $\mathfrak{F}(\tilde{\mu}) < \mathfrak{F}(\mu)$ which is a contradiction.

Case III.B. $\|\mathscr{H}^1 \llcorner \{\alpha_\eta(x) = r\} - \mu \llcorner \{\alpha_\eta(x) = r\}\| = \delta > 0\,.$
In this case, define

$$\tilde{\mu} := \mathscr{H}^1 \llcorner \{\alpha_\eta(x) \geq r\}\,.$$

One has that

$$C_{\tilde{\mu}}(\eta) = C_\mu(\eta) - r\delta, \text{ and } H(\|\tilde{\mu}\|) < H(\|\mu\|) + r\delta\,,$$

because $H$ is strictly concave and $\delta > 0$. This implies $\mathfrak{F}(\tilde{\mu}) < \mathfrak{F}(\mu)$ which is a contradiction. □

*Remark 5.29.* In the case when $H$ is strictly concave, applying to an arbitrary optimal measure $\mu$ both Theorems 5.21 and 5.28 one finds that $\mu = \mathscr{H}^1 \llcorner \Sigma$ with $\Sigma$ closed and $\mu = \mathscr{H}^1 \llcorner \Sigma'$ with $\Sigma'$ having countably many connected components. Clearly, one has then

$$\mathscr{H}^1(\Sigma \Delta \Sigma') = 0\,,$$

but $\Sigma$ and $\Sigma'$ could be different. Indeed, consider the classical Cantor "middle third" set

$$C := [0,1] \setminus \bigcup_{i=1}^{\infty} (a_i, b_i) \subseteq [0,1]\,,$$

where $(a_i, b_i)$ are the "middle third" intervals, and set

$$S := \bigcup_{i=1}^{\infty} (a_i, a_i + 8^{-i}) \cup (b_i - 8^{-i}, b_i)\,.$$

Then

$$\bar{S} = S \cup \bigcup_{i=1}^{\infty} \{a_i + 8^{-i}, b_i - 8^{-i}\} \cup C\,,$$

and hence, $\mathscr{H}^1(\bar{S}) = \mathscr{H}^1(S)$. Further, every singleton $\{x\}$ with

$$x \in C \setminus \bigcup\nolimits_i \{a_i, b_i\}$$

is a connected component of $\bar{S}$, and therefore, $S$ has more than countably many connected components. Now, in a way similar to Example 5.30 below, one can find two measures $f^+$ concentrated on the set

$$\bigcup_{i=1}^{\infty} \{a_i\} \times \{0\}$$

and $f^-$ concentrated on the set

$$\bigcup_{i=1}^{\infty} \{b_i\} \times \{0\}\,,$$

and a function $H$, such that the only optimal measure is given by

$$\mu = \mathscr{H}^1 \llcorner \Sigma = \mathscr{H}^1 \llcorner \bar{\Sigma}\,,$$

where $\Sigma := S \times \{0\}$. On the other hand, $\Sigma$ has countably many connected components but is not closed, and its closure $\bar{\Sigma} = \bar{S} \times \{0\}$ has more than countably many connected components.

The following example similar to Examples 5.22 and 5.23 shows that the result of the above Theorem 5.28 is optimal in the sense that in some situations no optimal set has finitely many connected components, even if the function $H$ is strictly concave.

*Example 5.30.* Let $\Omega \subseteq \mathbb{R}^2$ be a domain containing the segment $[0,1] \times \{0\}$, and define $f^+$ and $f^-$ as follows:

$$f^+ := \frac{1}{2}\delta_{(0,0)} + \sum_{n\geq 1} \frac{1}{2^{n+1}}\delta_{(2/3^n,0)} ,$$

$$f^- := \frac{1}{2}\delta_{(1,0)} + \sum_{n\geq 1} \frac{1}{2^{n+2}}\big(\delta_{(1/3^n,0)} + \delta_{(3/3^n,0)}\big) .$$

We will choose later a strictly concave function $H$ such that $H'(s) > 0$ for all $s > 0$. As in the previous examples it is easy to deduce that all the optimal measures must be supported in $[0,1] \times \{0\}$. Moreover, there is a t.p.m. which is optimal for all measures supported in $[0,1] \times \{0\}$, and hence for all optimal measures. Such a t.p.m. is given by the formula

$$\eta = \frac{1}{4}\,\delta_{[0,1]} + \sum_{n\geq 1} \frac{1}{2^{n+2}}\left(\delta_{[2\cdot 3^{-n},3\cdot 3^{-n}]} + \delta_{[2\cdot 3^{-n},1]} + \delta_{[0,3^{-n}]}\right),$$

where $\delta_{[p,q]}$ is the Dirac mass concentrated on the element of $\Theta$ given by the segment connecting $(p,0)$ to $(q,0)$ (note that for every measure $\mu$ supported in $[0,1] \times \{0\}$ there must be an optimal t.p.m. which moves $f^+$ along $[0,1] \times \{0\}$ from left to right, and that $\eta$ is the unique t.p.m. with the latter property). One can then write down the function $\alpha_\eta$, which is given by $\alpha_\eta(x,y) = 0$ for any $y \neq 0$, and

$$\alpha_\eta(x,0) = \begin{cases} \dfrac{1}{2} - \dfrac{1}{2^{n+2}} & \text{for } 1/3^n \leq x \leq 2/3^n , \\[2ex] \dfrac{1}{2} + \dfrac{1}{2^{n+2}} & \text{for } 2/3^n \leq x \leq 3/3^n . \end{cases} \tag{5.25}$$

Keeping in mind Remark 5.9, and in particular the formula (5.9) for $\beta$, we deduce that

$$\beta'(s) = \begin{cases} \dfrac{1}{2} + \dfrac{1}{2^{n+2}} & \text{for } \frac{1-1/3^{n-1}}{2} \leq s \leq \frac{1-1/3^n}{2} , \\[2ex] \dfrac{1}{2} - \dfrac{1}{2^{n+2}} & \text{for } \frac{1+1/3^n}{2} \leq s \leq \frac{1+1/3^{n-1}}{2} . \end{cases}$$

Notice that $\beta$ is concave according to Remark 5.9. We can now define $H$ setting $H(0) := 0$ and $H'(s)$ equal to

$$\begin{cases} \frac{1}{2} + \frac{1}{2^{n+2}} + \frac{3^n}{2^{n+2}}\left(\frac{1-1/3^n}{2} - x\right) & \text{for } \frac{1-1/3^{n-1}}{2} \le s \le \frac{1-1/3^n}{2}, \\ \frac{1}{2} - \frac{1}{2^{n+3}} - \frac{3^n}{2^{n+3}}\left(x - \frac{1+1/3^n}{2}\right) & \text{for } \frac{1+1/3^n}{2} \le s \le \frac{1+1/3^{n-1}}{2}. \end{cases}$$

According to the definition above, $H$ is a strictly concave function. Moreover, one has $H'(s) < \beta'(s)$ for all $0 \le s < 1/2$, while $H'(s) > \beta'(s)$ for all $s > 1/2$. Therefore, the function $s \mapsto H(s) - \beta(s)$ has a unique minimum in $s = 1/2$. Recalling Remark 5.9 and the fact that $\eta$ is an optimal t.p.m. with respect to any optimal measure $\mu$, we derive that the optimal measures are exactly the elements of $\mathcal{M}_{1/2}$. Finally, formula (5.25) for $\alpha_\eta$ ensures that the unique element of $\mathcal{M}_{1/2}$ is $\mu = \mathscr{H}^1 \llcorner \Sigma$, where

$$\Sigma := \left(\bigcup_{n\in\mathbb{N}} \left[\frac{2}{3^n}, \frac{3}{3^n}\right]\right) \times \{0\},$$

which clearly has countably many connected components. It is worth noting that $\overline{\Sigma} := \Sigma \cup \{0\}$ is still an optimal set having countably many connected components and is also closed. However, there is no optimal set with only a finite number of connected components, since for every Borel set

$$\Sigma' \subseteq [0,1] \times \{0\}$$

with such property one has $\mathscr{H}^1(\Sigma \Delta \Sigma') > 0$.

# Chapter 6
# Optimal Sets and Geodesics in the Two-Dimensional Case

This chapter is dedicated to the analysis of the two-dimensional case under the assumptions $A \equiv \mathrm{Id}$, $B \equiv 0$ and with no extra requirements on the function $H$. We will prove that in this case, if $f^+$, $f^- \in L^\infty(\Omega)$ (i.e. the measures $f^+$ and $f^-$ are absolutely continuous with respect to the Lebesgue measure on $\mathbb{R}^2$ and their densities, that we still denote by $f^+$ and $f^-$, belong to $L^\infty(\Omega)$), then every optimal set $\Sigma$ is contained in a finite number of geodesics with respect to $d_\Sigma$. Note that this result cannot be considered an improvement of Theorem 5.28, since we will not show $\Sigma$ to have finitely many connected components, but only to be contained in a finite union of connected paths. In fact, in Example 5.23 we have shown a situation in which the optimal set $\Sigma$ is contained in the single geodesic $[0,1] \times \{0\}$, but is totally disconnected.

**Theorem 6.1.** *If $A(t) = t$, $B(t) = 0$, $\Omega \subseteq \mathbb{R}^2$ and $f^\pm \in L^\infty(\Omega)$, then any optimal set $\Sigma$ is contained in a finite number of Lipschitz paths of bounded lengths, which are geodesics with respect to $\Sigma$. More precisely, there exists an ordered optimal t.p.m. $\eta$ such that $\Sigma$ is contained in finitely many paths $\theta \in \operatorname{spt} \eta$.*

To prove this result, we first fix an optimal set $\Sigma$ and denote $\mu := \mathscr{H}^1 \llcorner \Sigma$. Then, given any optimal ordered t.p.m. $\eta \in \mathcal{M}^+_{\mu,2}(\Theta)$, we recall that by Remark 5.6 there is a maximal constant $r > 0$ satisfying (5.2), so that one has in particular

$$\{\alpha_\eta(x) \geq r\} \subseteq \Sigma \subseteq \{\alpha_\eta(x) > r\}.$$

Our first goal is to show the following result.

**Proposition 6.2.** *With the above definitions one has $r > 0$.*

This is a crucial step to obtain Theorem 6.1. The proof of Proposition 6.2 is quite involved; we dedicate the whole next section to show that the assumption $r = 0$ leads to a contradiction.

G. Buttazzo et al., *Optimal Urban Networks via Mass Transportation*, Lecture Notes in Mathematics 1961, DOI: 10.1007/978-3-540-85799-0_6,

## 6.1 Preliminary Constructions

In this section we show that the hypothesis $r = 0$ leads to a contradiction. To do that, we perform a careful analysis of the consequences of this case. We stress that all the results of this section strongly use the (contradictory) assumption $r = 0$, hence they may not hold true in general. Recall that throughout this section we assume fixed an optimal t.p.m. $\eta \in \mathcal{M}^+_{\mu,2}(\Theta)$, so that $MK(\Sigma) = C_\Sigma(\eta)$.

**Lemma 6.3.** *Under the (contradictory) assumption $r = 0$, given a path $\sigma \in \operatorname{spt}\eta$ such that $\mu(\sigma) > 0$, and defining $x'$ and $y'$ as in Lemma 5.20, the whole open arc $\widetilde{x'y'}$ is contained in $\Sigma$.*

*Proof.* Recalling the proof of Lemma 5.20, we have that in particular

$$\alpha_\eta(z) > 0 \qquad\qquad \forall\, z \in \widetilde{x'y'}\,.$$

Indeed, for any such $z$, an $\eta$−positive quantity of paths close to $\sigma$ passes through $z$. By definition of $r$ and since $r = 0$, it follows that

$$z \in \Sigma \qquad\qquad \forall\, z \in \widetilde{x'y'}\,.$$

So the claim follows. □

*Remark 6.4.* We notice that $\Sigma$ is contained in countably many paths belonging to $\operatorname{spt}\eta$: this follows by Remark 5.8 since in the hypotheses of Theorem 6.1 condition (4.38) holds. Therefore, by Lemma 6.3 we infer that $\Sigma$ is made by countably many connected components $\{\Sigma_i\}_{i\in\mathbb{N}}$; in particular, any path in $\operatorname{spt}\eta$ may intersect at most one of the sets $\Sigma_i$.

It is well-known that $\mathscr{H}^1(\Sigma) = \mathscr{H}^1(\overline{\Sigma})$ whenever $\Sigma$ is a connected set. Hence, denoting by $\Sigma_i$, $i = 1, 2, \ldots$ the connected components of $\Sigma$, we may consider each $\Sigma_i$ to be closed. Define now

$$\begin{aligned} &\Theta_i := \{\theta \in \operatorname{spt}\eta \,:\, \theta \cap \Sigma_i \neq \emptyset\}, \qquad\qquad i = 1, 2, \ldots\,, \\ &\Theta_0 := \operatorname{spt}\eta \setminus \cup_i \Theta_i \end{aligned}$$

(observe that all $\Theta_i$, $i \neq 0$, are closed since so are assumed $\Sigma_i$). We obtain from Lemma 6.3 that all $\Theta_i$ are pairwise disjoint, so that one can write

$$\eta = \sum_{i=0}^{\infty} \eta_i, \text{ where } \eta_i := \eta \llcorner \Theta_i\,.$$

Note that each path $\theta \in \operatorname{spt}\eta_0$ is a line segment connecting $\theta(0)$ to $\theta(1)$ (since so is every $\theta \in \Theta_0$, because such a path by definition does not touch $\Sigma$ and is a geodesic). On the other hand, every path $\theta \in \operatorname{spt}\eta_i$ is made by three parts: the first one is a line segment connecting $\theta(0)$ to the point $s(\theta) \in \Sigma_i$ where $\theta$ touches $\Sigma$ for the first time, the second part is a path inside $\Sigma_i$, and the third is a line segment between a point of $\Sigma$ and $\theta(1)$. Notice that the function $s : \operatorname{spt} \eta_i \to \Sigma_i$ is well-defined and measurable. It is also worth observing that $s(\theta)$ is one of the points of $\Sigma_i$ of least distance from $\theta(0)$. We can then define a sort of "entry measure" by setting

$$\psi := s_{\#}\eta_i \in \mathcal{M}^+(\Sigma_i)\,.$$

Note that $\psi$ measures the mass entering inside $\Sigma_i$, and the above definition makes sense only thanks to the (contradictory) assumption $r = 0$: in fact, in the above construction, we used in a crucial way Lemma 6.3 which implies that each path of $\operatorname{spt} \eta$ may intersect only one of the connected components $\Sigma_i$ of $\Sigma$, while this property may be not true if $r > 0$.

**Lemma 6.5.** *The measure $\psi$ is non-atomic.*

*Proof.* Suppose by contradiction that there exists a $z \in \Sigma_i$ such that

$$\psi(\{z\}) > 0\,.$$

Since $r = 0$, we can take $\Sigma' \subseteq \Sigma$ such that $\mathscr{H}^1(\Sigma \setminus \Sigma') = \varepsilon$ and

$$\alpha_\eta(x) < \frac{\psi(\{z\})}{2(2\pi+1)} \qquad \text{for any } x \in \Sigma \setminus \Sigma' \tag{6.1}$$

(notice that we do not require $\Sigma \setminus \Sigma'$ to be contained in $\Sigma_i$). We define now the set $\Delta$, drawn in thick lines in Figure 6.1, as the union of the circle centered in $z$ of radius $\varepsilon/(2\pi+1)$ and a radial segment connecting the circle to $z$. Then $\mathscr{H}^1(\Delta) = \varepsilon$, so that $\Sigma'' := \Sigma' \cup \Delta$ has the same length as $\Sigma$. We introduce now a new t.p.m. $\tilde\eta := \rho_{\#}\eta$ with the following Borel function $\rho : \Theta \to \Theta$:

- if $\theta \in \operatorname{spt} \eta_i$, $s(\theta) = z$ and

$$|\theta(0) - z| \geq \frac{\varepsilon}{2\pi+1}\,,$$

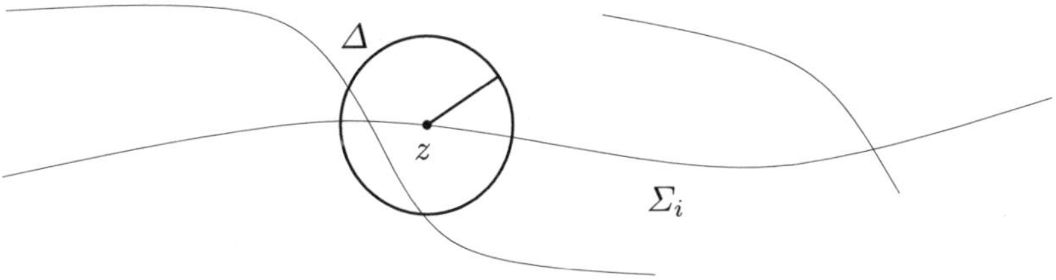


**Fig. 6.1** Definition of $\Delta$ in Lemma 6.5

then $\rho(\theta)$ is the path following $\theta$ until it reaches the circle, then moving inside $\Delta \subseteq \Sigma''$ to reach the point $z$, and finally following again $\theta$ from $z$ to $\theta(1)$;
- otherwise, $\rho(\theta) := \theta$.

Let us compute now $C_{\Sigma''}(\tilde{\eta})$: for any path $\theta$ such that $\rho(\theta) = \theta$ one has

$$\delta_{\Sigma''}\big(\rho(\theta)\big) \leq \delta_{\Sigma'}(\theta)\,,$$

while if $\theta$ belongs to the set

$$\Gamma := \{\theta : \rho(\theta) \neq \theta\}\,,$$

then

$$\delta_{\Sigma''}\big(\rho(\theta)\big) = \delta_{\Sigma'}(\theta) - \frac{\varepsilon}{2\pi + 1}\,.$$

To verify the latter equality, just recall that each $\theta \in \Gamma$ between $\theta(0)$ and $s(\theta) = z$ is a line segment. We have

$$\begin{aligned} C_{\Sigma''}(\tilde{\eta}) &= \int_\Theta \delta_{\Sigma''}(\theta)\, d\rho_\# \eta(\theta) = \int_\Theta \delta_{\Sigma''}\big(\rho(\theta)\big)\, d\eta(\theta) \\ &\leq \int_\Theta \delta_{\Sigma'}(\theta)\, d\eta(\theta) - \frac{\varepsilon}{2\pi+1}\eta(\Gamma) \\ &= C_{\Sigma'}(\eta) - \frac{\varepsilon}{2\pi+1}\eta(\Gamma)\,. \end{aligned} \tag{6.2}$$

Now we compare $C_\Sigma(\eta)$ and $C_{\Sigma'}(\eta)$. By definition of $\Sigma'$ and recalling (5.8) and (6.1), we obtain

$$\begin{aligned} C_{\Sigma'}(\eta) &= C_\emptyset(\eta) - \int_{\Sigma'} \alpha_\eta(x)\, d\mathscr{H}^1(x) \\ &= C_\emptyset(\eta) - \int_\Sigma \alpha_\eta(x)\, d\mathscr{H}^1(x) + \int_{\Sigma\setminus\Sigma'} \alpha_\eta(x)\, d\mathscr{H}^1(x) \\ &< C_\Sigma(\eta) + \frac{\psi(\{z\})}{2(2\pi+1)}\,\varepsilon\,. \end{aligned} \tag{6.3}$$

Finally, we need to estimate $\eta(\Gamma)$. By definition, we have

$$\eta(\Gamma) \geq \psi(\{z\}) - f^+\Big(B\big(z, \varepsilon/(2\pi+1)\big)\Big)\,,$$

and hence, since $f^+(\{z\}) = 0$, we can choose $\varepsilon$ small enough in such a way that

$$\eta(\Gamma) > \frac{\psi(z)}{2}\,.$$

This, together with (6.2), (6.3) and the fact that $\mathscr{H}^1(\Sigma) = \mathscr{H}^1(\Sigma'')$, gives

$$\begin{aligned}\mathfrak{F}(\Sigma'') &\le C_{\Sigma''}(\tilde\eta) + H\big(\mathscr{H}^1(\Sigma'')\big)\\ &\le C_{\Sigma'}(\eta) - \frac{\varepsilon}{2\pi+1}\,\eta(\Gamma) + H\big(\mathscr{H}^1(\Sigma)\big)\\ &< C_{\Sigma}(\eta) + \frac{\psi(\{z\})}{2(2\pi+1)}\,\varepsilon - \frac{\varepsilon}{2\pi+1}\,\eta(\Gamma) + H\big(\mathscr{H}^1(\Sigma)\big)\\ &< C_{\Sigma}(\eta) + H\big(\mathscr{H}^1(\Sigma)\big) = \mathfrak{F}(\Sigma)\,,\end{aligned}$$

which leads to a contradiction since $\mu = \mathscr{H}^1 \llcorner \Sigma$ was supposed to be an optimal measure. □

*Remark 6.6.* Notice that the proof of the above lemma did not use the hypothesis that the measures $f^+$, $f^-$ belong to $L^\infty$, but only that they be non-atomic.

We fix now arbitrarily two orthogonal axes $\hat e_1$ and $\hat e_2$ in $\mathbb{R}^2$. Let $k\colon \mathbb{R}^2 \to \Sigma_i$ stand for the Borel projection map onto $\Sigma_i$ (the latter is well-known to be defined for $\mathscr{L}^2$-a.e., hence for $f^\pm$-a.e. $x \in \mathbb{R}^2$, namely, for all $x \in \mathbb{R}^2$, for which the distance function $x \mapsto \operatorname{dist}(x, \Sigma_i)$ is not differentiable). Let $x \in \Sigma_i$ be such that the set $k^{-1}(x)$ belongs to some line $l$ and is not reduced to a point. Let then

$$\omega(x) \in [-\pi/2, \pi/2)$$

stand for the angle between $l$ and $\hat e_1$. We show now the following assertion.

**Lemma 6.7.** *The entry angle $\omega\colon \Sigma_i \to [-\pi/2, \pi/2)$ coincides $\psi$-a.e. with a Borel function.*

*Proof.* Consider a point $y \in \Sigma_i$: the set $K(y)$ of all the points $x \in \Omega$ such that

$$\operatorname{dist}(x, \Sigma_i) = |x - y|$$

is convex, since it is the intersection of the half-planes

$$\big\{x \in \Omega : |x-y| \le |x-z|\big\}$$

with $z \in \Sigma_i$. By construction, for any point $x$ in the interior $U(y)$ of $K(y)$, the point $y$ is the *unique* point of $\Sigma_i$ where $\operatorname{dist}(x, \Sigma_i)$ is attained; therefore, all the sets $U(y)$ are disjoint. Finally, for any $y \in \Sigma_i$, we have:

- either $K(y)$ is a segment of nonzero length, and then $\omega(y)$ is well-defined;
- or $U(y)$ has a strictly positive Lebesgue measure, which may happen for at most countably many points $y$ (and hence, on a $\psi$-negligible set by Lemma 6.5);
- or $K(y)$ reduces to a point, which happens only for such points $y \in \Sigma_i$, that if there is some $\theta \in \operatorname{spt}\eta$, $s(\theta) = y$, then $\theta(0) = y$ (recall that $\theta$ is a line segment between $\theta(0)$ and $s(\theta)$).

Therefore, the set $\Delta \subseteq \Sigma_i$ of the latter points has

$$\psi(\Delta) = f^+(\Delta) \leq f^+(\Sigma_i) = 0$$

since $f^+ \ll \mathscr{L}^2$. The Borel property for $\omega$ easily follows by taking a Borel selection $S : \Sigma_i \to \Theta_i$ such that

$$s(S(y)) = y \qquad\qquad \text{for } \psi - a.e.\ y \in \Sigma_i$$

and noticing that

$$\cos\omega(y) = \frac{\langle y - \big(S(y)\big)(0), \mathrm{e}_1\rangle}{\big|y - \big(S(y)\big)(0)\big|}$$

for $\psi$−a.e. $y$. □

In the sequel we will write $\omega(\theta)$ instead of $\omega(s(\theta))$. By the above lemma, in this way $\omega$ is defined for $\eta_i$-a.e. $\theta \in \Theta$. Summarizing, we know now that for $\psi$-a.e. point $z$ of $\Sigma_i$ there is a line passing through $z$ and containing $\theta(0)$ for each path $\theta \in \operatorname{spt}\eta_i$ such that $s(\theta) = z$. The angle between the oriented segment $\overrightarrow{\theta(0)s(\theta)}$ and the direction $\hat{e}_1$, which we further will call "oriented entry angle", must be either equal to $\omega(\theta)$ or to $\omega(\theta) + \pi$. We will then say that a path $\theta \in \operatorname{spt}\eta_i$ "enters in $\Sigma_i$ from the left" (resp. from the right), if the oriented entry angle associated to $\theta$ belongs to $[\pi/2, 3/2\pi)$ (resp. to $[-\pi/2, \pi/2)$). We will then write $\eta_i = \eta_l + \eta_r$ where $\eta$−a.e. path in the support of $\eta_l$ (resp. $\eta_r$) enters from the left (resp. from the right). As a consequence, we can write $\psi = \psi_l + \psi_r$, where

$$\psi_l := s_{\#}\eta_l\,, \qquad\qquad \psi_r := s_{\#}\eta_r\,.$$

We prove now the crucial property that $\psi_l = \psi_r$, that is for $\psi$−a.e. point of $\Sigma_i$ the mass entering from the left and the mass entering from the right are equal.

**Lemma 6.8.** *One has* $\psi_l = \psi_r$.

*Proof.* Fix an arbitrary injective path $\tau \subseteq \Sigma_i$ and any point $x \in \tau$. Letting $\bar{t}$ be the instant such that $\tau(\bar{t}) = x$ and considering $\tau$ as parametrized at speed one (that is, $|\tau'| \equiv 1$, at least in a neighborhood of $\bar{t}$), we denote by $x_\sigma$ the point $\tau(\bar{t} + \sigma)$, so that

$$\mathscr{H}^1\big(\widetilde{xx_\sigma}\big) = |\sigma|\,.$$

Choose now $\varepsilon > 0$, $N = N(\varepsilon) \in \mathbb{N}$ and $\delta = \delta(\varepsilon, N) > 0$ and change $\Sigma_i$ as follows. First of all, we cut the subpath $\widetilde{x_{-\varepsilon}x_\varepsilon}$ of $\tau$ joining $x_{-\varepsilon}$ and $x_\varepsilon$ and shift it leftwards (that is, in the direction of $-\hat{e}_1$) by $\delta$. Since in this way $\tau$ is divided in three parts, to keep the connectedness of $\tau$ we add two horizontal segments of length $\delta$. Moreover, we notice that in this way $\Sigma_i$ can become disconnected because the affluents entering in $\tau$ between $x_{-\varepsilon}$ and $x_\varepsilon$ (*a priori*,

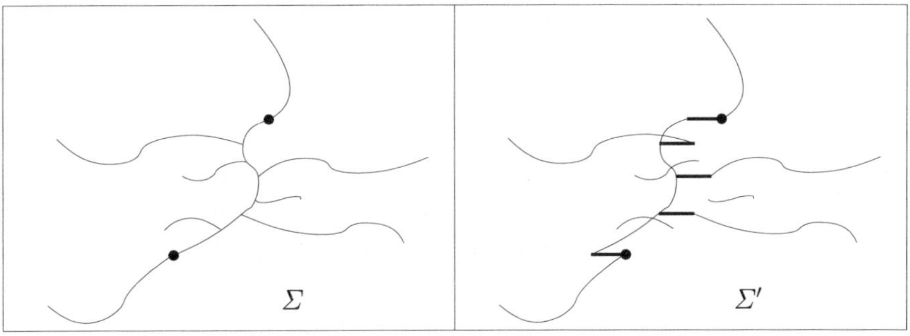


**Fig. 6.2** Construction of $\Sigma'$ from $\Sigma$ in Lemma 6.8

there could be countably many of these affluents) may remain disconnected after this change. Hence, we choose $N-2$ of these affluents, namely those carrying more mass, and we connect them too with the new path $\tau$, by means of $N-2$ horizontal segments of length $\delta$. In this way, starting from $\Sigma$ we have built a new set $\Sigma'$ satisfying

$$\mathscr{H}^1(\Sigma') \leq \mathscr{H}^1(\Sigma) + N\delta\,.$$

Figure 6.2 illustrates the above construction (the $N$ segments added in $\Sigma'$ are drawn darker than the rest of the figure). We want to compare now $MK(\Sigma')$ to $MK(\Sigma)$. To do that we use an argument similar to the one of Lemma 6.5, but a bit more involved: we choose an optimal t.p.m. $\eta \in \mathcal{M}^+_{\mu,2}(\Theta)$, so that

$$MK(\Sigma) = C_\Sigma(\eta)$$

and we construct the new t.p.m. $\eta' = \rho_\#\eta$, where the map $\rho$: $\operatorname{spt}\eta \to \Theta$ is defined as follows. First of all, if $\theta$ does not intersect the arc $\widetilde{x_{-\varepsilon}x_\varepsilon}$, then simply $\rho(\theta) = \theta$, and hence

$$\delta_{\Sigma'}\big(\rho(\theta)\big) \leq \delta_\Sigma\big(\theta\big)\,. \tag{6.4}$$

Otherwise, we decompose $\theta = \theta_1\cdot\theta_2\cdot\theta_3$, where $\theta_1$ is the part of $\theta$ between $\theta(0)$ and the first point of contact with the arc $\widetilde{x_{-\varepsilon}x_\varepsilon}$, $\theta_2$ is the part of $\theta$ between its first and last point of contact with the arc $\widetilde{x_{-\varepsilon}x_\varepsilon}$, and, finally, $\theta_3$ is the part of $\theta$ between its last point of contact with the arc $\widetilde{x_{-\varepsilon}x_\varepsilon}$ and $\theta(1)$. We write then $\rho(\theta) := \tilde\theta_1\cdot\tilde\theta_2\cdot\tilde\theta_3$, where the paths $\tilde\theta_i$ are defined as follows. The path $\tilde\theta_2$ is the leftwards shift by $\delta$ of the part of the arc $\widetilde{x_{-\varepsilon}x_\varepsilon}$ between $\theta_2(0)$ and $\theta_2(1)$, so that

$$\delta_{\Sigma'}\big(\tilde\theta_2\big) = \delta_\Sigma\big(\theta_2\big) = 0\,. \tag{6.5}$$

Concerning $\theta_1$, there are four possibilities.

**(i)** If $\theta$ enters in $\widetilde{x_{-\varepsilon}x_\varepsilon}$ from $x_{-\varepsilon}$ or $x_\varepsilon$, then $\tilde\theta_1$ is defined as the composition of $\theta_1$ with a horizontal segment of length $\delta$ connecting $\tilde\theta_1$ to $\tilde\theta_2$, and therefore

$$\delta_{\Sigma'}\big(\tilde\theta_1\big) = \delta_\Sigma\big(\theta_1\big)\,. \tag{6.6}$$

**(ii)** If $\theta$ enters in $\widetilde{x_{-\varepsilon}x_\varepsilon}$ from one of the biggest $N-2$ affluents, then we let $\tilde\theta_1$ to be again just the composition of $\theta_1$ with a horizontal segment of length $\delta$, and hence by construction of $\Sigma'$ one still has

$$\delta_{\Sigma'}\big(\tilde\theta_1\big) = \delta_\Sigma\big(\theta_1\big)\,. \tag{6.7}$$

**(iii)** If $\theta$ enters in $\widetilde{x_{-\varepsilon}x_\varepsilon}$ from an affluent which is not among the biggest $N-2$, then we let $\tilde\theta_1$ to be still $\theta_1$ composed with a horizontal segment of length $\delta$ connecting $\theta_1$ to $\tilde\theta_2$, but this time the segment we added is not inside $\Sigma'$, and therefore

$$\delta_{\Sigma'}\big(\tilde\theta_1\big) \le \delta_\Sigma\big(\theta_1\big) + \delta\,. \tag{6.8}$$

**(iv)** If $\theta$ enters in $\Sigma_i$ directly in the open arc $\widetilde{x_{-\varepsilon}x_\varepsilon}$, then $\theta_1$ is a line segment with endpoints $\theta(0)$ and $s(\theta)=\theta_2(0)$. In this case we simply let $\tilde\theta_1$ be the line segment between $\theta(0)$ and $\tilde\theta_2(0)$. By a straightforward computation we can compare the lengths of $\theta_1$ and $\tilde\theta_1$ by means of the entry angle $\omega$ (recall that the latter is defined in $s(\theta)$ for $\eta$-a.e. $\theta$): in fact,

$$\begin{aligned}\delta_{\Sigma'}\big(\tilde\theta_1\big) &\le \overline{\theta(0)\tilde\theta_2(0)} = \overline{\theta(0)\theta_2(0)} \pm \delta\cos\omega\big(s(\theta)\big) + \delta o_\theta(1)\\ &= \delta_\Sigma\big(\theta_1\big) \pm \delta\cos\omega\big(s(\theta)\big) + \delta o_\theta(1)\,,\end{aligned} \tag{6.9}$$

where the sign is "$-$" if $\theta$ enters from the left and "$+$" if $\theta$ enters from the right, and where for any $\theta$ the quantity $o_\theta(1)$ vanishes as $\delta\to 0$ (not necessarily uniformly in $\theta$) and $|o_\theta(1)|\le 1$.

The way to define $\tilde\theta_3$ is completely symmetric.

Recalling the definition of $\rho(\theta)$ and the estimates (6.4)–(6.9), one can immediately notice that

$$\begin{aligned} MK\big(\Sigma'\big) &\le C_{\Sigma'}(\eta')\\ &\le C_\Sigma(\eta) + \delta\Big(M_N - \int_{\widetilde{x_{-\varepsilon}x_\varepsilon}} \cos\omega(x)d(\psi_l-\psi_r)(x) + \int_{\Theta_{\rm iv}} o_\theta(1)\,d\eta\Big)\\ &= MK\big(\Sigma\big) + \delta\Big(M_N - \int_{\widetilde{x_{-\varepsilon}x_\varepsilon}} \cos\omega(x)d(\psi_l-\psi_r)(x) + \int_{\Theta_{\rm iv}} o_\theta(1)\,d\eta\Big)\,.\end{aligned}$$

Here $M_N$ stands for the mass of the paths entering in $\widetilde{x_{-\varepsilon}x_\varepsilon}$ from the "small" affluents, plus the mass of the paths exiting from them, $\Theta_{\rm iv}$ is the set of paths as in **(iv)**, and the last equality is due to the fact that $MK(\Sigma)=C_\Sigma(\eta)$ in view of the optimality of $\eta$. We claim that

$$\lambda := \int_{\widetilde{x_{-\varepsilon}x_\varepsilon}} \cos\omega(x) d(\psi_l - \psi_r)(x) \le 0 \,. \tag{6.10}$$

Indeed, assume by contradiction that $\lambda > 0$; by construction, for any fixed $\varepsilon$, one has $M_N \to 0$ as $N \to \infty$, so we may fix an integer $N$ large enough such that $M_N < \lambda/3$. Moreover, since by the Dominated Convergence Theorem

$$\int_{\Theta_{\text{iv}}} o_\theta(1)\, d\eta$$

tends to zero as $\delta \to 0$, we may take $\delta$ small enough so that

$$\int_{\Theta_{\text{iv}}} o_\theta(1)\, d\eta < \lambda/3 \,.$$

Therefore

$$MK(\Sigma') \le MK(\Sigma) - C\delta$$

for some constant $C > 0$ which depends only on $\varepsilon$. Summarizing, if (6.10) is not true then starting from the set $\Sigma$ we would be able to find, for $\delta$ arbitrarily small a set $\Sigma'$ satisfying

$$\mathscr{H}^1(\Sigma') \le \mathscr{H}^1(\Sigma) + N\delta \,, \qquad MK(\Sigma') \le MK(\Sigma) - C\delta$$

with suitable constants $N$, $C > 0$. But this leads to conclude that $r \ge C/N$, which is impossible since we are assuming $r = 0$. In fact, if $r < C/N$, then one can find, for $\delta$ sufficiently small, a subset

$$\Delta \subseteq \Sigma' \cap \{\alpha_\eta < C/N\}$$

of length $N\delta$, so that by (5.8) for the set $\Sigma'' := \Sigma' \setminus \Delta$ one would have $MK(\Sigma'') < MK(\Sigma)$ while $\mathscr{H}^1(\Sigma'') \le \mathscr{H}^1(\Sigma)$, against the optimality of $\Sigma$. Therefore, (6.10) holds.

In the very similar way, moving the arc $\widetilde{x_{-\varepsilon}x_\varepsilon}$ rightwards rather than leftwards we conclude the opposite inequality, and therefore

$$\int_{\widetilde{x_{-\varepsilon}x_\varepsilon}} \cos\omega(x) d(\psi_l - \psi_r)(x) = 0 \,.$$

Analogously, moving the arc upwards and downwards, we obtain

$$\int_{\widetilde{x_{-\varepsilon}x_\varepsilon}} \sin\omega(x) d(\psi_l - \psi_r)(x) = 0 \,.$$

This concludes the proof since the arc $\widetilde{x_{-\varepsilon}x_\varepsilon}$ was arbitrarily chosen in $\Sigma$. □

**Lemma 6.9.** *There exists a constant $C$ such that for all sufficiently small $\varepsilon$ there exists a set $\Delta$ satisfying*

$$\mathscr{H}^1(\Delta) = \varepsilon \,, \qquad MK(\Sigma \cup \Delta) \le MK(\Sigma) - C\varepsilon^2 \,.$$

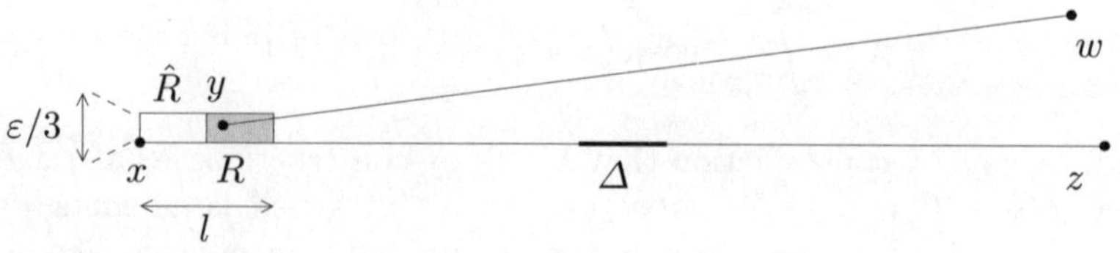


**Fig. 6.3** Construction in Lemma 6.9

*Proof.* Since $f^+(\Sigma) = 0$, there exists a connected component $\Sigma_i$ of the optimal set $\Sigma$ and a point $x \in \operatorname{spt}(\pi_{0\#}\eta) \setminus \Sigma_i$, where $\eta$ is an optimal t.p.m. associated to $\Sigma$. Take a path $\theta \in \operatorname{spt}\eta_i$ with $\theta(0) = x$, and denote

$$z := s(\theta) \in \Sigma_i .$$

Consider now a thin rectangle $\widehat{R}$ with sides $l \ll \overline{xz}$ and $\varepsilon/3 \ll l$ having a corner in $x$ and a long side contained in the segment $xz$. Divide then $\widehat{R}$ in two rectangles of sides $l/2$ and $\varepsilon/3$ and let $R$ be the right one: in Figure 6.3, the rectangle $\widehat{R}$ is the wide one, while $R$ is its shaded right half. Let now

$$\Sigma' := \Sigma \cup \Delta ,$$

where $\Delta = zw$ is a segment of length $\varepsilon$ contained in the middle of the segment $xz$.

Take now

$$y \in R \cap \operatorname{spt}(\pi_{0\#}\eta_i) ,$$

a path $\tau \in \operatorname{spt}\eta_i$ with $\tau(0) = y$, and let $w := s(\tau)$. By the optimality of $\eta$ and since $\Sigma_i$ is connected, we already know that $\overline{xz} \le \overline{xw}$ as well as $\overline{yw} \le \overline{yz}$. Denote by $y'$ the projection of $y$ onto $xz$. Then, whenever $l$ is small enough with respect to $\overline{xz}$, and $\varepsilon$ is sufficiently small so that $y'$ belongs to the segment $xz$ and does not belong to $\Delta$, one has the series of estimates

$$\begin{aligned}
d_{\Sigma'}(y,z) &\le \frac{\varepsilon}{3} + |y'-z| - \varepsilon \\
&= \frac{\varepsilon}{3} + |x-z| - |x-y'| - \varepsilon \\
&= |x-z| - |x-y'| - \frac{2\varepsilon}{3} \\
&\le |x-w| - |x-y'| - \frac{2\varepsilon}{3} \\
&\le |x-y| + |y-w| - |x-y'| - \frac{2\varepsilon}{3} \\
&\le |x-y'| + \frac{\varepsilon}{3} + |y-w| - |x-y'| - \frac{2\varepsilon}{3} \\
&= |y-w| - \frac{\varepsilon}{3} = d_\Sigma(y,w) - \frac{\varepsilon}{3} .
\end{aligned}$$

To conclude, it suffices now to notice that the area of $R$ is proportional to $\varepsilon$ (since $x$, $\theta$ and $l \ll \overline{xz}$ have been fixed), and, moreover, up to a suitable choice of $x$, $\theta$ and $l$, we can assume that $\pi_{0\#}\eta_i(R) \geq K\mathscr{L}^2(R)$ with a small constant $K$ not depending on $\varepsilon$ (here we use the assumption $f \in L^\infty$). We define $\eta' := \rho_\# \eta$, where $\rho(\theta) := \theta$ unless $\theta \in \operatorname{spt}\eta_i$ and $\theta(0) \in R$, in which case $\rho(\theta)$ replaces the line segment $\theta(0)s(\theta)$ by the composition of three segments $\theta(0)z$, $zw$ and $ws(\theta)$. The conclusion now follows because

$$\begin{aligned} MK(\Sigma \cup \Delta) &\leq C_{\Sigma\cup\Delta}(\eta') \leq C_\Sigma(\eta) - K\mathscr{L}^2(R)\,\frac{\varepsilon}{3} \\ &= MK(\Sigma) - \frac{Kl}{18}\,\varepsilon^2 . \end{aligned} \tag{6.11}$$

□

*Remark 6.10.* An inspection of the proof of Lemma 6.9 shows that the constant $C$ can be estimated from below as

$$C \geq \frac{f^+(x)\mathrm{dist}(x,\Sigma)}{C_0}$$

where $C_0$ is a geometric constant depending only on $\operatorname{diam}\Omega$ and $x$ is any Lebesgue point of $f^+$.

Finally, we can find a contradiction using Proposition 6.11 below, which will show that the assumption $r = 0$ cannot hold true and hence conclude the proof of Proposition 6.2.

**Proposition 6.11.** *There exists a set* $\widetilde{\Sigma}$ *such that*

$$\mathscr{H}^1(\widetilde{\Sigma}) = \mathscr{H}^1(\Sigma), \qquad\qquad MK(\widetilde{\Sigma}) < MK(\Sigma)$$

*(so that* $\mathfrak{F}(\widetilde{\Sigma}) < \mathfrak{F}(\Sigma)$ *against the optimality of* $\Sigma$*).*

*Proof.* Thanks to Lemma 5.15, $\Sigma_i$ can be endowed with the partial order generated by the relation given by $x \leq y \in \Sigma_i$ whenever there is a path $\theta \in \operatorname{spt}\eta_i$ passing through $x$ and, afterwards, through $y$. Therefore, for almost all $\varepsilon$ small enough (almost all is intended with respect to the Lebesgue measure on $\mathbb{R}$), it is possible to find a point $x_\varepsilon$ such that

$$\mathscr{H}^1(X_\varepsilon) = \varepsilon \qquad \text{where} \quad X_\varepsilon := \big\{x \in \Sigma_i :\ x \leq x_\varepsilon\big\}, \tag{6.12}$$

so that

$$\big\{x \in \Sigma_i :\ x \leq x_\varepsilon\big\}$$

is contained in the ball of center $x_\varepsilon$ and radius $\varepsilon$. By Lemma 6.7, again for almost all $\varepsilon$ we know that the entry angle $\omega(x_\varepsilon)$ is well-defined. Moreover, as a consequence of Lemma 6.9, we may prove that at least a mass $C\varepsilon$ passes through $X_\varepsilon$: in other words,

$$\psi(X_\varepsilon) = s_\# \eta_i(X_\varepsilon) \geq C\varepsilon\,, \tag{6.13}$$

the constant $C$ being that of Lemma 6.9, thus independent of $\varepsilon$. Indeed, applying Lemma 6.9 find the set $\Delta$, and denote by $\eta'$ the t.p.m. built in the proof of the lemma: by (6.11) we know that

$$C_{\Sigma\cup\Delta}(\eta') \leq MK(\Sigma) - C\varepsilon^2\,.$$

If we now set

$$\Sigma' := \big(\Sigma \cup \Delta\big) \setminus X_\varepsilon\,,$$

which satisfies $\mathscr{H}^1(\Sigma') \leq \mathscr{H}^1(\Sigma)$, then by construction of $\eta'$ one clearly has

$$C_{\Sigma'}(\eta') \leq C_{\Sigma\cup\Delta}(\eta') + \varepsilon\psi(X_\varepsilon)\,.$$

Hence, by the optimality of $\Sigma$ we conclude

$$\begin{aligned} MK(\Sigma) \leq MK(\Sigma') \leq C_{\Sigma'}(\eta') &\leq C_{\Sigma\cup\Delta}(\eta') + \varepsilon\psi(X_\varepsilon) \\ &\leq MK(\Sigma) - C\varepsilon^2 + \varepsilon\psi(X_\varepsilon)\,, \end{aligned}$$

thus (6.13) is established. Let us now set

$$K := \sqrt{\frac{C}{8\pi\|f^+\|_{L^\infty}}}\,.$$

Since $f^+\big(B(x_\varepsilon, K\sqrt{\varepsilon})\big) \leq C\varepsilon/8$, while by (6.13)

$$\begin{aligned} &f^+\big(\{x \in \Omega : \exists\theta \in \operatorname{spt}\eta_i, \theta(0) = x,\, s(\theta) \in X_\varepsilon\}\big) \\ &\qquad\qquad \geq \eta\big(\{\theta \in \operatorname{spt}\eta_i : s(\theta) \in X_\varepsilon\}\big) = \psi(X_\varepsilon) \geq C\varepsilon\,, \end{aligned}$$

recalling also Lemma 6.8 we deduce that $\varepsilon$ can be chosen with the additional property that there are two paths $\theta_P$ and $\theta_Q$ in $\operatorname{spt}\eta_i$ entering in $\Sigma_i$ respectively from the left and from the right in such a way that

$$s(\theta_P) = s(\theta_Q) = x_\varepsilon$$

and both the points $P = \theta_P(0)$ and $Q = \theta_Q(0)$ are outside the ball $B\big(x_\varepsilon, K\sqrt{\varepsilon}\big)$. For simplicity, we assume that $x_\varepsilon \equiv (0,0)$ and that $\omega(x_\varepsilon) = 0$.

Since each path enters in $\Sigma_i$ at a point of least distance from its first extreme, $\Sigma_i$ intersects neither the open ball $B(P, \overline{Px_\varepsilon})$ nor the open ball $B(Q, \overline{Qx_\varepsilon})$. Thus, $X_\varepsilon$ is constrained in the shaded region of Figure 6.4. We claim now that the entry angle $\omega(x)$ is "small" for most of the points $x \in X_\varepsilon$; more precisely one has

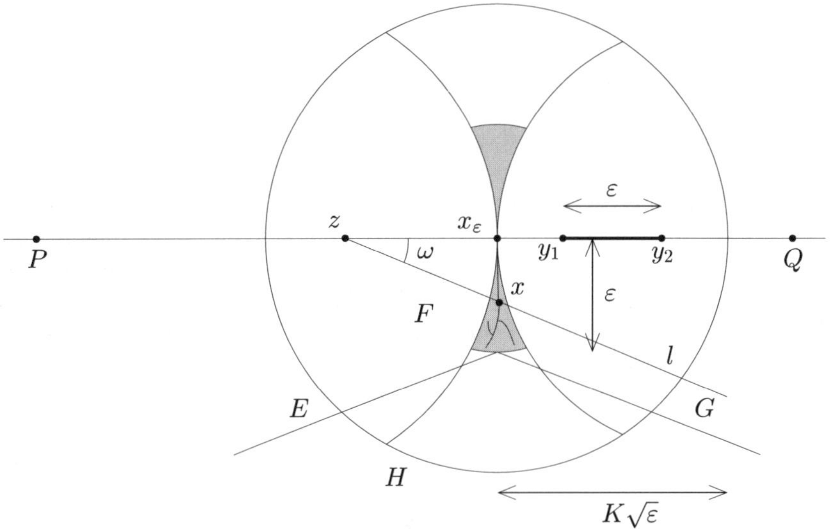


**Fig. 6.4** Situation in Proposition 6.11

$$\psi\Big(\Big\{x \in X_\varepsilon : \ |\omega(x)| > 2\,\frac{\sqrt{\varepsilon}}{K}\Big\}\Big) \leq \frac{C}{8}\,\varepsilon\,. \tag{6.14}$$

In fact, take a point $x \in X_\varepsilon$ such that

$$|\omega(x)| > 2\frac{\sqrt{\varepsilon}}{K} \ :$$

we will say that $x \in X_L$ if $x$ is above $x_\varepsilon$ and $\omega(x) > 0$ or if $x$ is below $x_\varepsilon$ and $\omega(x) < 0$ (the figure shows this second possibility), while we will say that $x \in X_R$ otherwise. By construction, if $x \in X_L$ (resp. $x \in X_R$) then the line $l$ passing through $x$ with slope $\omega(x)$ intersects the horizontal axis at a point $z$ in the interior of the ball

$$B\left(x_\varepsilon, K\sqrt{\frac{\varepsilon}{2}}\,\right),$$

and $z$ is in the left (resp. right) of $x_\varepsilon$. Since for any $\theta \in \operatorname{spt}\eta_i$, $s(\theta)$ is the point of $\Sigma_i$ closest to $\theta(0)$, the segment $Px_\varepsilon$ cannot intersect the segment $\theta(0)x$ whenever $\theta \in \operatorname{spt}\eta_i$ and $\theta(0) = x$. For any such $\theta$, the point $\theta(0)$ belongs to the line $l$ by definition of the entry angle $\omega(x)$, therefore if $x \in X_L$ (resp. $x \in X_R$) and $\theta$ enters at $x$ from the left (resp. from the right), then $\theta(0)$ belongs to the segment $zx$, thus to the ball $B(x_\varepsilon, K\sqrt{\varepsilon})$. We have then

$$\begin{aligned}\psi_l(X_L) + \psi_r(X_R) &= \eta_i\Big(\big\{\theta \in \operatorname{spt}\eta_l : s(\theta) \in X_L\big\} \cup \big\{\theta \in \operatorname{spt}\eta_r : s(\theta) \in X_R\big\}\Big)\\ &\leq f^+\Big(B\Big(x_\varepsilon, K\sqrt{\varepsilon/2}\Big)\Big) \leq \frac{C}{16}\,\varepsilon\,.\end{aligned}$$

Recalling that $\psi_l = \psi_r$ by Lemma 6.8, we deduce

$$\begin{aligned}\psi(X_\varepsilon) &= \psi_l(X_L) + \psi_r(X_L) + \psi_r(X_R) + \psi_l(X_R)\\ &\leq 2\big(\psi_l(X_L) + \psi_r(X_R)\big) \leq \frac{C}{8}\,\varepsilon\end{aligned}$$

and thus (6.14) is proved.

In order to define the competitor $\widetilde{\Sigma}$ to $\Sigma$, we divide $\mathbb{R}^2$ in four regions; since these regions are symmetric with respect to the horizontal axis, we draw only the bottom half in Figure 6.4 to avoid confusion. The region $F$ is the ball $B(x_\varepsilon, K\sqrt{\varepsilon})$; the region $H$ is made by such points $\xi \notin F$ below (resp. above) the horizontal axis, such that the angle between the segment connecting $\xi$ to the point $(0,-\varepsilon)$ (resp. $(0,\varepsilon)$) and the horizontal axis has the absolute value greater than $\sqrt{\varepsilon}/2K$. The regions $E$ and $G$ are the left and the right side of the remaining part of $\mathbb{R}^2$. For simplicity of notation we denote by $m(E)$ the value

$$\eta\big(\{\theta \in \eta_i :\ \theta(0) \in E\}\big)\,,$$

and we adopt the analogous notation for the regions $F$, $G$ and $H$. Further, without loss of generality, we assume $m(G) \geq m(E)$.

We now define the competitor $\widetilde{\Sigma}$: consider the segment $y_1y_2$ (drawn in bold in the figure), where

$$y_1 \equiv \Big(\frac{K}{2}\sqrt{\varepsilon} - \frac{\varepsilon}{2}, 0\Big)\,, \qquad\qquad y_2 \equiv \Big(\frac{K}{2}\sqrt{\varepsilon} + \frac{\varepsilon}{2}, 0\Big)\,;$$

we let then

$$\widetilde{\Sigma} := \big(\Sigma \setminus X_\varepsilon\big) \cup y_1y_2\,.$$

By (6.12), we have $\mathscr{H}^1\big(\widetilde{\Sigma}\big) \leq \mathscr{H}^1\big(\Sigma\big)$, so we only need to check that $MK\big(\widetilde{\Sigma}\big) < MK\big(\Sigma\big)$ to find the desired contradiction.

As in the last lemmas, to show that $MK\big(\widetilde{\Sigma}\big) < MK\big(\Sigma\big)$ we will provide a suitable t.p.m. $\eta' := \rho_\# \eta$ with $C_{\widetilde{\Sigma}}(\eta') < C_\Sigma(\eta)$. In order to construct the map $\rho : \Theta \to \Theta$, we first set $\rho(\theta) := \theta$ for every

$$\theta \notin \Theta_\varepsilon := \{\theta \in \operatorname{spt} \eta_i,\ s(\theta) \in X_\varepsilon\}\,.$$

For $\theta \notin \Theta_\varepsilon$, of course we have $\delta_{\widetilde{\Sigma}}\big(\rho(\theta)\big) \leq \delta_\Sigma(\theta)$. Concerning now the case $\theta \in \Theta_\varepsilon$, if $\theta(0) \in F$ we again let $\rho(\theta) := \theta$, so that

$$\theta(0) \in F \Longrightarrow \delta_{\widetilde{\Sigma}}\big(\rho(\theta)\big) \leq \delta_\Sigma(\theta) + \varepsilon\,. \tag{6.15}$$

If $\theta(0) \in H$, we still let $\rho(\theta) := \theta$ and by construction, we have

$$\theta(0) \in H \Longrightarrow \delta_{\widetilde{\Sigma}}\big(\rho(\theta)\big) \leq \delta_\Sigma(\theta) + \varepsilon\,. \tag{6.16}$$

We call now

$$\hat{t} = \hat{t}(\theta) := \max\{t : \theta(t) \in X_\varepsilon\}.$$

If $\theta(0) \in E$, we set $\rho(\theta)$ to be the path obtained joining the line segment $\theta(0)\theta(\hat{t})$ to the curve $\theta \llcorner [\hat{t}, 1]$. By definition of the set $E$, a simple geometric estimates ensures

$$\begin{aligned}\theta(0) \in E \Longrightarrow \delta_{\widetilde{\Sigma}}\big(\rho(\theta)\big) &\leq \delta_\Sigma(\theta) + \overline{\theta(0)\theta(\hat{t})} - \overline{\theta(0)s(\theta)} \\ &\leq \delta_\Sigma(\theta) + \varepsilon \sin\big(\sqrt{\varepsilon}/2K\big) + O(\varepsilon^{3/2}) \\ &= \delta_\Sigma(\theta) + o(\varepsilon),\end{aligned} \tag{6.17}$$

where $o(\varepsilon)$ depends only on $\varepsilon$ and not on $\theta$. Finally, if $\theta(0) \in G$, we define $\rho(\theta)$ as the path joining the line segments $\theta(0)y_2$, $y_2y_1$, and $y_1\theta(\hat{t})$ to the curve $\theta \llcorner [\hat{t}, 1]$. Another easy geometric argument gives

$$\theta(0) \in G \Longrightarrow \delta_{\widetilde{\Sigma}}\big(\rho(\theta)\big) \leq \delta_\Sigma(\theta) - \varepsilon + O(\varepsilon^{3/2}) \leq \delta_\Sigma(\theta) - \varepsilon + o(\varepsilon). \tag{6.18}$$

Now, (6.15)–(6.18) together yield

$$\begin{aligned}MK\big(\widetilde{\Sigma}\big) \leq MK\big(\Sigma\big) &+ m(F)\,\varepsilon + m(H)\,\varepsilon \\ &+ m(E)\,o(\varepsilon) - m(G)\,\varepsilon + m(G)\,o(\varepsilon).\end{aligned} \tag{6.19}$$

By (6.13), we know that

$$m(E) + m(F) + m(G) + m(H) \geq C\varepsilon.$$

By (6.14), we know that

$$m(H) \leq \frac{1}{8}\,C\varepsilon;$$

by definition, we know that

$$m(F) \leq \frac{1}{8}\,C\varepsilon.$$

Hence, since we are assuming $m(G) \geq m(E)$, we find

$$m(G) \geq \frac{3}{8}\,C\varepsilon.$$

Therefore, (6.19) gives us that

$$MK\big(\widetilde{\Sigma}\big) \leq MK\big(\Sigma\big) + \varepsilon\left(\frac{1}{4}C - \frac{3}{8}C + Co(1)\right),$$

and then, provided $\varepsilon$ was chosen sufficiently small, we get $MK\big(\widetilde{\Sigma}\big) < MK\big(\Sigma\big)$, hence the required contradiction that completes the proof. □

Now, it is immediate to conclude the validity of Proposition 6.2

*Proof (of Proposition 6.2).* If $r = 0$, then the claims of Lemmas 6.3, 6.5, 6.7, 6.8, 6.9 and of Proposition 6.11 hold true; hence, there is a set $\widetilde{\Sigma}$ with $\mathfrak{F}(\widetilde{\Sigma}) < \mathfrak{F}(\Sigma)$, against the optimality of $\Sigma$. □

## 6.2 Proof of the Main Result

We can finally give the proof of Theorem 6.1, which makes use of the technical results of Section 5.2 and of the fact that $r > 0$ for any $\eta \in \mathcal{M}^+_{\mu,2}(\Theta)$ which we have shown in the previous section.

To prove Theorem 6.1 we need to select a special t.p.m. $\eta \in \mathcal{M}^+_2(\Omega)$. To this aim, let us first arbitrarily fix a function $\alpha : \Sigma \to \mathbb{R}^+$ such that there exists some $\eta \in \mathcal{M}^+_{\mu,2}(\Theta)$ for which $\alpha_\eta = \alpha$. We solve then another auxiliary maximization problem.

**Lemma 6.12.** *Defining* $F_1(\eta) := \max\{\mu(\theta), \theta \in \operatorname{spt}\eta\}$, *we have that the set*

$$\mathcal{M}^+_{\mu,3,1}(\Theta) := \operatorname{Arg\,max}\Big\{F_1(\eta) : \ \eta \in \mathcal{M}^+_{\mu,2}(\Theta), \ \alpha_\eta = \alpha\Big\} \tag{6.20}$$

*is non-empty and convex.*

We recall that the functional $F_1(\eta)$ is well-defined for all $\eta \in \mathcal{M}^+_{\mu,2}(\Theta)$ by Corollary 5.12.

*Proof (of Lemma 6.12).* Consider a maximizing sequence $\{\eta_n\} \subseteq \mathcal{M}^+_{\mu,2}(\Theta)$ for the functional $F_1$ under the constraint $\alpha_{\eta_n} = \alpha$, and define

$$\eta := \sum_{n\in\mathbb{N}} 2^{-n}\,\eta_n\,.$$

By Corollary 5.12, there is a path $\theta_n$ maximizing the length

$$\theta \mapsto \mu(\theta) = \mathscr{H}^1(\theta \cap \Sigma)$$

within $\operatorname{spt}\eta_n$. Since the Euclidean length of the paths $\theta_n$ is bounded thanks to Lemma 4.18 and since $\Omega$ is bounded, up to a subsequence we can assume that $\theta_n \to \theta$ for some path $\theta \in \Theta$. Notice that, since $\operatorname{spt}\eta$ is closed and contains $\theta_n$ by construction, we obtain $\theta \in \operatorname{spt}\eta$. Since we already know that $\theta \mapsto \mu(\theta)$ is an u.s.c. function, we obtain

$$\limsup_{n\to\infty} F_1(\eta_n) = \limsup_{n\to\infty} \mu(\theta_n) \le \mu(\theta) \le F_1(\eta)\,.$$

Therefore, $\eta \in \mathcal{M}^+_{\mu,3,1}(\Theta)$ and we have proved that the set $\mathcal{M}^+_{\mu,3,1}(\Theta)$ is non-empty. The convexity of this set follows immediately by noticing that a convex combination of elements of $\mathcal{M}^+_{\mu,3,1}(\Theta)$ is a measure, the support of which contains all the supports of the elements. □

Notice that the existence of solutions for the maximization problem (6.20) cannot be shown in the same way as done for $\mathcal{M}^+_{\mu,1}(\Theta)$ and $\mathcal{M}^+_{\mu,2}(\Theta)$ in Remark 5.11, where one had to deal with integral functionals and therefore it was sufficient to check the lower (or upper) semicontinuity of the integrands. Recall that in those cases, moreover, the standard direct methods of the calculus of variations also gave the weak* closedness of $\mathcal{M}^+_{\mu,1}(\Theta)$ and $\mathcal{M}^+_{\mu,2}(\Theta)$. On the contrary, it is fundamental to notice that (in general) $\mathcal{M}^+_{\mu,3,1}(\Theta)$ *is not weakly* closed*; indeed, if $\eta_n \overset{*}{\rightharpoonup} \eta$, $\theta_n \in \operatorname{spt}\eta_n$ and $\theta_n \to \theta$, then in general it is not true that $\theta \in \operatorname{spt}\eta$. For this reason, to extract a particular t.p.m. in $\mathcal{M}^+_{\mu,3,1}(\Theta)$ in the sequel we will not take a weak* limit of some optimizing sequence, since it may happen that this limit does not belong to $\mathcal{M}^+_{\mu,3,1}(\Theta)$; instead, we will repeat the method already used in the proof of Lemma 6.12, which only needs the convexity of $\mathcal{M}^+_{\mu,3,1}(\Theta)$ in place of the weak* closedness.

In the sequel, we will often use a t.p.m. $\eta \in \mathcal{M}^+_{\mu,3,1}(\Theta)$ together with a path $\theta_1$ providing the maximum in $F_1$; therefore, it is useful to set

$$\Pi_1 := \Big\{(\eta,\theta_1) :\ \eta \in \mathcal{M}^+_{\mu,3,1}(\Theta),\ \mu(\theta_1) = F_1(\eta)\}\Big\}.$$

Let us now prove a "second-order" version of Corollary 5.12.

**Lemma 6.13.** *For any* $(\eta,\theta_1) \in \Pi_1$ *there exists a* $\tau \in \Theta$ *maximizing*

$$\tau \mapsto \mathscr{H}^1 \llcorner (\Sigma \setminus \theta_1)(\tau)$$

*within* $\operatorname{spt}\eta$.

*Proof.* Fix $(\eta,\theta_1) \in \Pi_1$, take a sequence $\{\tau_n\}$ maximizing

$$\tau \mapsto \mathscr{H}^1 \llcorner (\Sigma \setminus \theta_1)(\tau)$$

within $\operatorname{spt}\eta$, and assume without loss of generality (thanks to Lemma 4.18) that $\tau_n \to \tau$ in $\Theta$. As in Corollary 5.12, we conclude applying Lemma 4.1 with $X := \Theta$, $C_n := \theta_n$ and $\nu = \tilde\mu := \mathscr{H}^1 \llcorner (\Sigma\setminus\theta_1)$; indeed, the lemma gives

$$\tilde\mu(\tau) \geq \limsup_{n\to\infty} \tilde\mu(\tau_n),$$

and this clearly concludes the proof. □

We define

$$F_2(\eta,\theta_1) := \max\{\mathscr{H}^1 \llcorner (\Sigma \setminus \theta_1)(\tau),\ \tau \in \operatorname{spt}\eta\};$$

$$\widetilde{\mathcal{M}}^+_{\mu,3,2}(\Theta) := \operatorname{Arg\,max}\Big\{F_2(\eta,\theta_1),\ (\eta,\theta_1) \in \Pi_1\Big\};$$

$$\widetilde{\Pi}_2 := \Big\{(\eta,\theta_1,\tilde\theta_2) :\ (\eta,\theta_1) \in \widetilde{\mathcal{M}}^+_{\mu,3,2}(\Theta),$$

$$\tilde{\theta}_2 \in \operatorname{Arg\,max}\{\mathscr{H}^1 \llcorner (\Sigma \setminus \theta_1)(\tau),\, \tau \in \operatorname{spt}\eta\}\Big\};$$

$$\Pi_2 := \operatorname{Arg\,max}\Big\{\mu(\theta_2),\ (\eta,\theta_1,\theta_2) \in \widetilde{\Pi}_2\Big\};$$

$$\mathcal{M}^+_{\mu,3,2}(\Theta) := \Big\{\eta \in \mathcal{M}^+_{\mu,3,1}(\Theta) :\ \exists\theta_1,\, \theta_2 \in \Theta, (\eta,\theta_1,\theta_2) \in \Pi_2\Big\}.$$

We underline the meaning of the sets we defined. Recall that our aim is to cover $\Sigma$ with paths, possibly in an efficient way; therefore, we defined $\mathcal{M}^+_{\mu,3,1}(\Theta)$ to be the set of the optimal measures $\eta \in \mathcal{M}^+_{\mu,2}(\Theta)$, having $\alpha_\eta$ fixed, the longest path of which inside $\Sigma$ has the maximal possible length. The elements of $\Pi_1$ are the pairs of such particular optimal measures and their paths using $\Sigma$ as much as possible. To proceed covering $\Sigma$, we find now the "second longest path", namely the path in $\operatorname{spt}\eta$ having the maximal length inside $\Sigma \setminus \theta_1$. Hence, $\widetilde{\mathcal{M}}^+_{\mu,3,2}(\Theta)$ are the pairs $(\eta,\theta_1) \in \Pi_1$ maximizing this new length that is possible to gain by this second longest path, and $\widetilde{\Pi}_2$ are the corresponding triples given by the pairs

$$(\eta,\theta_1) \in \widetilde{\mathcal{M}}^+_{\mu,3,2}(\Theta)$$

together with a second path $\tilde{\theta}_2$ maximizing the new length. For reasons that will become clear during the proof of Theorem 6.1, among those triples we have the convenience to make a further choice: for all the triples

$$(\eta,\theta_1,\theta_2) \in \widetilde{\Pi}_2$$

the length

$$\mathscr{H}^1 \llcorner (\Sigma \setminus \theta_1)(\theta_2)$$

is the same, and it is the maximal possible; we then select those triples for which

$$\mu(\theta_2) = \mathscr{H}^1 \llcorner \Sigma(\theta_2)$$

is maximal. In words, among the different paths maximizing the new length that one can gain by $\theta_2$, we choose those maximizing also the total length $\mu(\theta_2)$. Those particular triples are collected in $\Pi_2$, and finally $\mathcal{M}^+_{\mu,3,2}(\Theta)$ consists of the optimal measures $\eta$ which belong to $\Pi_2$ together with some pair of paths $(\theta_1,\, \theta_2) \in \operatorname{spt}\eta$.

We now extend Lemma 6.12 to $\mathcal{M}^+_{\mu,3,2}(\Theta)$.

**Lemma 6.14.** *The set* $\mathcal{M}^+_{\mu,3,2}(\Theta)$ *is non-empty and convex.*

*Proof.* We first show that $\widetilde{\mathcal{M}}^+_{\mu,3,2}(\Theta)$ is non-empty: let $(\eta_n,\theta_{1,n})$ be a maximizing sequence for the functional $F_2$ in $\Pi_1$. In particular, according to Lemma 6.13, one has

$$F_2(\eta_n,\theta_{1,n}) = \mathscr{H}^1 \llcorner (\Sigma \setminus \theta_{1,n})(\tilde{\theta}_{2,n})$$

for some $\tilde{\theta}_{2,n} \in \operatorname{spt} \eta_n$. Then, as in Lemma 6.12, we define

$$\eta := \sum_{n\in\mathbb{N}} 2^{-n}\eta_n \,,$$

and we let $\theta_1$ and $\tilde{\theta}_2$ be the limits in $\Theta$ of $\theta_{1,n}$ and $\tilde{\theta}_{2,n}$ respectively (up to a subsequence). We claim that

$$(\eta, \theta_1, \tilde{\theta}_2) \in \widetilde{\Pi}_2 \,, \tag{6.21}$$

hence in particular $\widetilde{\mathcal{M}}^+_{\mu,3,2}(\Theta)$ contains $(\eta, \theta_1)$, so it is not empty.

To show (6.21), we notice that $\eta \in \mathcal{M}^+_{\mu,2}(\Theta)$ and $\alpha_\eta = \alpha$ as in Lemma 6.12; moreover,

$$\mu(\theta_1) \geq \limsup_{n\to\infty} \mu(\theta_{1,n})$$

by upper semicontinuity (recall Lemma 4.1). Then, since in this case all the paths $\theta_{1,n}$ have the same, maximal, length in $\Sigma$, we deduce that the above inequality is actually an equality, so that $(\eta, \theta_1) \in \Pi_1$. To show (6.21), we only need to prove that

$$\limsup_{n\to\infty} \mathscr{H}^1 \llcorner (\Sigma \setminus \theta_{1,n})(\tilde{\theta}_{2,n}) \leq \mathscr{H}^1 \llcorner (\Sigma \setminus \theta_1)(\tilde{\theta}_2) \,,$$

which is equivalent to

$$\limsup_{n\to\infty} \mu\big(\tilde{\theta}_{2,n} \setminus \theta_{1,n}\big) - \mu\big(\tilde{\theta}_2 \setminus \theta_1\big) \leq 0 \,.$$

In view of the inclusion

$$(A \setminus B) \setminus (C \setminus D) \subseteq (A \setminus C) \cup (D \setminus B) \,,$$

one has

$$\begin{aligned} \mu\big(\tilde{\theta}_{2,n} \setminus \theta_{1,n}\big) - \mu\big(\tilde{\theta}_2 \setminus \theta_1\big) &\leq \mu\Big(\big(\tilde{\theta}_{2,n} \setminus \theta_{1,n}\big) \setminus \big(\tilde{\theta}_2 \setminus \theta_1\big)\Big) \\ &\leq \mu\big(\tilde{\theta}_{n,2} \setminus \tilde{\theta}_2\big) + \mu\big(\theta_1 \setminus \theta_{n,1}\big) \,, \end{aligned} \tag{6.22}$$

so that it suffices to show that the lim sup of the right hand-side is zero. Concerning $\mu\big(\tilde{\theta}_{n,2} \setminus \tilde{\theta}_2\big)$, it is immediately seen that its lim sup is zero thanks to Lemma 4.1 with $\nu = \mathscr{H}^1 \llcorner \big(\Sigma \setminus \tilde{\theta}_2\big)$. In fact,

$$\mu\big(\tilde{\theta}_{n,2} \setminus \tilde{\theta}_2\big) = \nu(\tilde{\theta}_{n,2})$$

and of course $\nu(\tilde{\theta}_2) = 0$. Concerning $\mu\big(\theta_1 \setminus \theta_{n,1}\big)$, by the equality

$$\big|\mu(A) - \mu(B)\big| = \big|\mu(A \setminus B) - \mu(B \setminus A)\big| \,,$$

valid for any measure $\mu$ and sets $A$ and $B$, one has

$$\left|\mu(\theta_1) - \mu(\theta_{n,1})\right| = \left|\mu(\theta_1 \setminus \theta_{n,1}) - \mu(\theta_{n,1} \setminus \theta_1)\right|.$$

As we already noticed,

$$\mu(\theta_{n,1}) = \mu(\theta_1) \qquad \forall n \in \mathbb{N};$$

moreover,

$$\mu(\theta_{n,1} \setminus \theta_1) \to 0$$

exactly as previously done for $\tilde{\theta}_2$ and $\tilde{\theta}_{n,2}$. Then, it follows that also

$$\mu(\theta_1 \setminus \theta_{n,1}) \to 0$$

so that the lim sup of the right hand-side in (6.22) is zero, thus (6.21) follows: as noticed before, this gives that $\widetilde{\mathcal{M}}^+_{\mu,3,2}(\Theta)$ is non-empty.

Moreover, $\widetilde{\mathcal{M}}^+_{\mu,3,2}(\Theta)$ is also convex, as one can immediately deduce exactly as done for $\mathcal{M}^+_{\mu,3,1}(\Theta)$ at the end of Lemma 6.12.

Concerning $\mathcal{M}^+_{\mu,3,2}(\Theta)$, the convexity can be deduced again in the very same way as for $\mathcal{M}^+_{\mu,3,1}(\Theta)$ and $\widetilde{\mathcal{M}}^+_{\mu,3,2}(\Theta)$. To show that $\mathcal{M}^+_{\mu,3,2}(\Theta)$ is non-empty, we will check that so is $\Pi_2$. In fact, it suffices again to take a sequence

$$\left\{(\eta_n, \theta_{1,n}, \theta_{2,n})\right\} \subseteq \widetilde{\Pi}_2$$

such that $\theta_{2,n}$ is a maximizing sequence for $\mu$. Then define, possibly up to a subsequence,

$$\eta := \sum_{n\in\mathbb{N}} 2^{-n}\eta_n$$

and $\theta_1$ and $\theta_2$ to be the limits in $\Theta$ of $\theta_{1,n}$ and $\theta_{2,n}$ respectively. By the same arguments as before we have

$$(\eta, \theta_1, \theta_2) \in \widetilde{\Pi}_2\,,$$

and moreover

$$\mu(\theta_2) \geq \limsup \mu(\theta_{2,n})$$

by upper semicontinuity again. Then $\Pi_2 \neq \emptyset$ as $(\eta, \theta_1, \theta_2) \in \Pi_2$ and the proof is concluded. □

We define now by induction, for any $n \in \mathbb{N}$,

$$F_n(\eta, \theta_1, \ldots, \theta_{n-1}) := \max\{\mathscr{H}^1 \llcorner \left(\Sigma \setminus (\theta_1 \cup \cdots \cup \theta_{n-1})\right)(\tau),\ \tau \in \operatorname{spt}\eta\}$$

$$\widetilde{\mathcal{M}}^+_{\mu,3,n}(\Theta) := \operatorname{Arg\,max}\Big\{F_n(\eta, \theta_1, \ldots, \theta_{n-1}) :\ (\eta, \theta_1, \ldots, \theta_{n-1}) \in \Pi_{n-1}\Big\};$$

$$\widetilde{\Pi}_n := \Big\{(\eta, \theta_1, \ldots \theta_{n-1}, \tilde{\theta}_n) :\ (\eta, \theta_1, \ldots, \theta_{n-1}) \in \widetilde{\mathcal{M}}^+_{\mu,3,n}(\Theta),$$

$$\mathscr{H}^1 \llcorner \big(\Sigma \setminus (\theta_1 \cup \dots \cup \theta_{n-1})\big)(\tilde{\theta}_n) = F_n(\eta, \theta_1, \dots, \theta_{n-1})\Big\};$$

$$\Pi_n := \operatorname{Arg\,max}\Big\{\mu(\theta_n) : \ (\eta, \theta_1, \dots, \theta_n) \in \widetilde{\Pi}_n\Big\};$$

$$\mathcal{M}^+_{\mu,3,n}(\Theta) := \Big\{\eta \in \mathcal{M}^+_{\mu,3,n-1} : \ \exists \theta_1, \dots, \theta_n \in \Theta : \ (\eta, \theta_1, \dots, \theta_n) \in \Pi_n\Big\}.$$

Further, set

$$\Pi_\infty := \Big\{(\eta, \theta_1, \dots, \theta_n, \dots) : \ \forall n \in \mathbb{N}, \ (\eta, \theta_1, \dots, \theta_n) \in \mathcal{M}^+_{\mu,3,n}\Big\};$$

$$\mathcal{M}^+_{\mu,3,\infty}(\Theta) := \Big\{\eta \in \mathcal{M}^+_{\mu,2}(\Theta) : \ \exists \theta_1, \dots, \theta_n, \dots \in \Theta : \\ (\eta, \theta_1, \dots, \theta_n, \dots) \in \Pi_\infty\Big\}.$$

We have the following result.

**Lemma 6.15.** *All the sets* $\mathcal{M}^+_{\mu,3,n}(\Theta)$ *are non-empty and convex, as well as* $\mathcal{M}^+_{\mu,3,\infty}(\Theta)$.

*Proof.* The first part of the statement can be obtained exactly as in Lemma 6.14 via an inductive procedure. Concerning the second part, it can also be obtained in a similar way: take a sequence $\{\eta_n\} \in \mathcal{M}^+_{\mu,2}(\Theta)$ with $\alpha_{\eta_n} = \alpha$ and paths $\theta_{m,n} \in \Theta$ with $m \leq n$ such that for any $n \in \mathbb{N}$ one has

$$(\eta_n, \theta_{1,n}, \dots, \theta_{n,n}) \in \Pi_n\,.$$

By a standard diagonal argument, we can assume that for any $m \in \mathbb{N}$ one has $\theta_{m,n} \to \theta_m$ as $n \to \infty$, and we define as usual

$$\eta := \sum_{n \in \mathbb{N}} 2^{-n} \eta_n\,;$$

then, arguing as in the above lemmas we derive that

$$(\eta, \theta_1, \dots, \theta_m) \in \mathcal{M}^+_{\mu,3,m} \qquad\qquad \forall\, m \in \mathbb{N}\,,$$

so that

$$(\eta, \theta_1, \dots, \theta_n, \dots) \in \Pi_\infty\,.$$

Hence, the fact that $\mathcal{M}^+_{\mu,3,\infty}(\Theta)$ is non-empty is established. The convexity of $\mathcal{M}^+_{\mu,3,\infty}(\Theta)$ is straightforward. Indeed, for any

$$\eta_1, \ \eta_2 \in \mathcal{M}^+_{\mu,3,\infty}(\Theta)$$

and any $0 < t < 1$ one has

$$\operatorname{spt}\big(t\eta_1 + (1-t)\eta_2\big) \supseteq \operatorname{spt}\eta_1\,.$$

Thus, if $(\eta, \theta_1, \ldots, \theta_n, \ldots) \in \Pi_\infty$, one has also

$$(t\eta_1 + (1-t)\eta_2, \theta_1, \ldots, \theta_n, \ldots) \in \Pi_\infty$$

by definition of $\Pi_\infty$. □

*Proof (of Theorem 6.1).* Take an optimal set $\Sigma$ and a t.p.m. $\eta \in \mathcal{M}^+_{\mu,3,\infty}(\Theta)$ with

$$(\eta, \theta_1, \ldots, \theta_n, \ldots) \in \Pi_\infty$$

for a suitable sequence $\{\theta_n\} \subseteq \operatorname{spt} \eta$. The thesis will be achieved once shown that

$$\Sigma \subseteq \theta_1 \cup \theta_2 \cup \cdots \cup \theta_n$$

for some $n \in \mathbb{N}$. If this is not true, then one has $\mu(\theta_i) > 0$ for all $i \in \mathbb{N}$. We define now $x_i$ and $y_i$ those points of $\theta_i$ such that $\widetilde{x_i y_i}$ is the shortest subpath of $\theta_i$ with $\mu(\widetilde{x_i y_i}) = \mu(\theta_i)$. Now, the proof follows in five steps.

Step I. *For any $i \in \mathbb{N}$, there is at least a mass $r > 0$ passing through $x_i$ (resp. $y_i$) and following the path $\theta_i$ for a while after $x_i$ (resp. before $y_i$).*
Fix $i \in \mathbb{N}$, and define

$$\Theta_i := \Big\{\theta \in \operatorname{spt}\eta : \ \exists t_1 < t_2,\, t_1' < t_2',\, \theta(t_1) = x_i, \\ \theta(t_2) \neq x_i,\, \theta \llcorner [t_1, t_2] = \theta_i \llcorner [t_1', t_2']\Big\},$$

$$\widetilde{\Theta}_i := \Big\{\theta \in \operatorname{spt}\eta : \ \exists t_1 < t_2,\, t_1' < t_2',\, \theta(t_2) = y_i, \\ \theta(t_1) \neq y_i,\, \theta \llcorner [t_1, t_2] = \theta_i \llcorner [t_1', t_2']\Big\},$$

so that $\Theta_i$ are precisely those paths passing through $x_i$ and then following $\theta_i$ for at least a while, and $\widetilde{\Theta}_i$ are the paths passing through $y_i$ having followed $\theta_i$ for a while. The claim of this step is then $\eta(\Theta_i) \geq r$ and $\eta(\widetilde{\Theta}_i) \geq r$.

By definition of $x_i$ and $y_i$, we have a sequence $\{z_n\}$ of points contained in $\widetilde{x_i y_i}$, converging to $x_i$ and belonging to $\Sigma$. By (5.2), one has $\alpha_\eta(z_n) \geq r$, which can be rewritten as

$$\eta(\Theta_{z_n}) \geq r\,, \tag{6.23}$$

where

$$\Theta_z := \{\theta \in \Theta : \ z \in \theta\}$$

for any $z \in \Omega$. Suppose now by contradiction that

$$\eta(\Theta_i) = r - \delta$$

for some $\delta > 0$; then, by (6.23) one has

$$\eta(\Theta_{z_n} \setminus \Theta_i) > \delta \qquad \forall n \in \mathbb{N}\,. \tag{6.24}$$

We remark now that, by construction and thanks to Lemma 5.20,

$$\Theta_i = \bigcup\nolimits_{j\in\mathbb{N}} \bigcap\nolimits_{n\geq j} \Theta_{z_n} \tag{6.25}$$

up to negligible sets. Moreover, again by Lemma 5.20,

$$\{\Theta_{z_1} \cap \Theta_{z_n}\}$$

is an $\eta-$essentially decreasing sequence (more precisely, the sequence of sets $\Theta_{z_1} \cap \Theta_{z_n} \cap \operatorname{spt} \eta$ is decreasing). Therefore, also the sequence

$$\{(\Theta_{z_1} \cap \Theta_{z_n}) \setminus \Theta_i\}$$

is $\eta-$essentially decreasing and its intersection must be $\eta-$negligible by (6.25). This implies the existence of an integer $n > 1$ such that

$$\eta\big((\Theta_{z_1} \cap \Theta_{z_n}) \setminus \Theta_i\big) < \eta\big(\Theta_{z_1} \setminus \Theta_i\big) - \delta\,.$$

So we assume without loss of generality (up to redefining the sequence $\{z_n\}$) that

$$\eta\big((\Theta_{z_1} \cap \Theta_{z_2}) \setminus \Theta_i\big) < \eta\big(\Theta_{z_1} \setminus \Theta_i\big) - \delta\,.$$

(the right hand-side of the above inequality is a positive number by (6.24). Applying (6.24) to $n = 2$ gives

$$\eta\big((\Theta_{z_1} \cup \Theta_{z_2}) \setminus \Theta_i\big) = \eta\big(\Theta_{z_2} \setminus \Theta_i\big) + \eta\big(\Theta_{z_1} \setminus \Theta_i\big) - \eta\big((\Theta_{z_1} \cap \Theta_{z_2}) \setminus \Theta_i\big) > 2\delta\,.$$

Arguing in the same way, for any $n \in \mathbb{N}$ we can obtain

$$\eta\Big((\Theta_{z_1} \cup \Theta_{z_2} \cup \cdots \cup \Theta_{z_n}) \setminus \Theta_i\Big) > n\delta\,.$$

Since of course this is not possible for $n > \|\eta\|/\delta$, the contradiction shows that $\eta(\Theta_i) \geq r$; in the very same way, $\eta(\widetilde{\Theta}_i) \geq r$).

<u>Step II.</u> *If $x_i \neq x_j$ (resp. $y_i \neq y_j$), one has $\Theta_i \cap \Theta_j = \emptyset$ (resp. $\widetilde{\Theta}_i \cap \widetilde{\Theta}_j = \emptyset$).* Assume that this is not true, so take $\theta \in \Theta_i \cap \Theta_j$, with $j > i$ and with $x_i \neq x_j$. Suppose that $\theta$ passes first through $x_j$, than through $x_i$. Then, since $\theta \in \Theta_i$ there is an interval $[t_1, t_2]$ such that

$$\theta(t_1) = \theta_i(t_1) = x_i$$

and

$$\theta \llcorner [t_1, t_2] \equiv \theta_i \llcorner [t_1, t_2]\,.$$

Moreover, since $\theta$ has passed through $x_j$ before passing through $x_i$ (and recalling that $\theta \in \Theta_j$) we have

$$\mu\big(\theta([0, t_1])\big) > 0 = \mu\big(\theta_i([0, t_1])\big)$$

by construction. We aim now to find a competitor to $\eta$ proving that $\eta \notin \Pi_i$, finding then a contradiction. The idea is to modify $\theta$ and $\theta_i$: more precisely, we define

$$\tilde{\theta} := \theta \llcorner [0, t_1] \cdot \theta_i \llcorner [t_1, 1]$$

as the path which follows $\theta$ from $\theta(0)$ to $x_i$, and then follows $\theta_i$ from $x_i$ to $\theta_i(1)$, and analogously we define

$$\tilde{\theta}_i := \theta_i \llcorner [0, t_1] \cdot \theta \llcorner [t_1, 1]$$

as the path following $\theta_i$ from $\theta_i(0)$ to $x_i$ and $\theta$ from $x_i$ to $\theta(1)$. Then, we may define a modification $\tilde{\eta}$ of $\eta$ as follows: we take an $\eta-$positive quantity of mass contained in a neighborhood of $\theta_i$ in $\Theta$ and passing through $x_i$ (this is possible thanks to Step I), and an equal mass contained in a neighborhood of $\theta$ in $\Theta$ and passing through $x_i$ (this is possible by Corollary 5.14 and Lemma 5.17). As in the proof of Lemma 5.15 we define $\tilde{\eta}$ swapping the ways of the paths when they meet at $x_i$, and leaving all the other paths unchanged. Therefore, $\tilde{\theta}$ and $\tilde{\theta}_i$ belong to $\operatorname{spt} \tilde{\eta}$. It is immediately seen that

$$C_\Sigma(\tilde{\eta}) = C_\Sigma(\eta) ,$$

so that $\tilde{\eta}$ is still an optimal t.p.m. Moreover, $\theta_1, \theta_2, \ldots, \theta_{i-1}$ belong to $\operatorname{spt} \tilde{\eta}$, provided we have chosen the neighborhoods of $\theta$ and $\theta_i$ in the construction above small enough. Recall also that

$$\Sigma \cap \theta_i \subsetneq \Sigma \cap \tilde{\theta} :$$

therefore, we obtain that

$$(\tilde{\eta}, \theta_i, \theta_2, \ldots, \theta_{i-1}, \tilde{\theta}) \in \widetilde{\Pi}_i :$$

moreover, since $\mu(\tilde{\theta}) > \mu(\theta_i)$, we have a contradiction with the assumption

$$(\eta, \theta_i, \theta_2, \ldots, \theta_{i-1}, \theta_i) \in \Pi_i .$$

Summarizing, we finally found a contradiction assuming that $\theta$ passes through $x_j$ before than through $x_i$.

If $\theta$ passes through $x_i$ before passing through $x_j$, a completely similar argument gives again a contradiction at the $j-$th step: in fact, an analogous change of the t.p.m. allows to assume that $\operatorname{spt} \eta$ contains the path

$$\tilde{\theta} := \theta \llcorner [0, t_1] \cdot \theta_j \llcorner [t_1, 1]$$

which follows $\theta$ between $\theta(0)$ and $x_j$, and $\theta_j$ between $x_j$ and $\theta_j(1)$. Therefore, at the $j-$th stage, it would be more convenient to select $\tilde{\theta}$ instead of $\theta_j$, since

$$\Sigma \cap \tilde{\theta} \supsetneq \Sigma \cap \theta_j :$$

we again found a contradiction, so that the property

$$\Theta_i \cap \Theta_j = \emptyset$$

is proved. Concerning

$$\widetilde{\Theta}_i \cap \widetilde{\Theta}_j = \emptyset\,,$$

the argument is of course exactly the same. Then also this step is concluded.

*Step III. If $x_i = x_j$ (resp. $y_i = y_j$), then either $\Theta_i \cap \Theta_j = \emptyset$ or $\Theta_i = \Theta_j$ (resp. either $\widetilde{\Theta}_i \cap \widetilde{\Theta}_j = \emptyset$ or $\widetilde{\Theta}_i = \widetilde{\Theta}_j$).*

Also for this step, we will only proof the first case, being the second completely similar. To achieve the step, we will show that if $\Theta_i \cap \Theta_j \neq \emptyset$, then $\theta_i$ and $\theta_j$ coincide for a while after $x_i$; formally, this means that if $x_i = x_j$ and $\Theta_i \cap \Theta_j \neq \emptyset$ then one has

$$\theta_i([s,t]) = \theta_j([s',t']) \supsetneq \{x_i\} \tag{6.26}$$

with

$$\theta_i(s) = \theta_j(s') = x_i = x_j\,.$$

This will immediately conclude the step by definition of $\Theta_i$ and $\Theta_j$, so it is enough to establish (6.26). To this aim, suppose that it is false: then there must exist a path $\bar\theta \in \Theta_i \cap \Theta_j$, and $\bar\theta$ should pass twice from $x_i$: once in order to follow $\theta_i$ for a while, and one to follow $\theta_j$ for another while. But this would mean that the path $\bar\theta$ contains a loop, which is impossible thanks to Remark 5.16. Notice that here it is crucial that the claims of Lemma 5.15 and of Remark 5.16 have been established for all the paths $\theta \in \operatorname{spt}\eta$, and not just for $\eta-$almost all.

*Step IV. If $\Theta_i = \Theta_j$ and $\widetilde{\Theta}_i = \widetilde{\Theta}_j$ (and so $x_i = x_j$, $y_i = y_j$ by Step II), then $\theta_i$ and $\theta_j$ coincide between $x_i$ and $y_i$.*

By Step III we already know that, for suitable $t_h$, $t'_h$, $1 \leq h \leq 4$, one has

$$\begin{aligned} &\theta_i \llcorner [t_1,t_2] = \theta_j \llcorner [t'_1,t'_2]\,, && \theta_i \llcorner [t_3,t_4] = \theta_j \llcorner [t'_3,t'_4]\,,\\ &\theta_i(t_1) = x_i, && \theta_i(t_4) = y_i\,. \end{aligned}$$

On the other hand, applying twice Lemma 5.20 we know that

$$\theta_i \llcorner [t_2,t_3] = \sigma_{\theta_i(t_2)\theta_i(t_3)} = \sigma_{\theta_j(t'_2)\theta_j(t'_3)} = \theta_j \llcorner [t'_2,t'_3]\,,$$

and so $\theta_i$ and $\theta_j$ coincide in the whole part between $x_i$ and $y_i$.

*Step V. Conclusion.*

Since $\|\eta\| = 1$, Steps I, II and III ensure that there can be at most $1/r$ different sets $\Theta_i$ and at most $1/r$ different sets $\widetilde{\Theta}_i$, hence at most $1/r^2$ different pairs $(\Theta_i, \widetilde{\Theta}_i)$. Since $\Sigma$ is trivially contained in the countable union of the arcs $\widetilde{x_i y_i}$, and since by Step IV these arcs are in fact a finite number, we have concluded the proof. □

# Appendix

## A The Mass Transportation Problem

The mass transportation problem was first proposed by G. Monge in 1781 [49] as follows: *"Étant donnés dans l'espace, deux volumes egaux entr'eux, & terminés chacune par une ou plusieurs surfaces courbes donnés; trouver dans le second volume le point où doit être transportée chaque molécule du premier, pour que la somme des produits des molecules multipliées chacune par l'espace parcouru soit un minimum"*. In modern language, we can rewrite his question as follows: given two sets $A$, $B \subseteq \mathbb{R}^3$ with the same volume, we look for a measurable map $T : A \to B$ which describes a way to transport the set $A$ onto $B$, and so that the cost

$$\int_A |T(x) - x| \, dx \tag{A.1}$$

is minimal. Saying that $T$ "describes a transportation" means that, by moving the mass on $A$ according to $T$, we completely cover $B$, that is, for any measurable set $E \subseteq B$ one must have

$$\text{meas}\Big(\Big\{x \in A : T(x) \in E\Big\}\Big) = \text{meas}(E) \,. \tag{A.2}$$

A first generalization that can be done is to consider a density of the material which may be non-constant or even singular; then, instead of two sets, we can consider two positive Borel measures $f^+$ and $f^-$ over a Polish space $X$ which have the same total mass, that is $\|f^+\| = \|f^-\|$. Usually, one assumes this total mass to be unitary for simplicity. A Borel map $T : X \to X$ is then called a *transport map* if $T_\# f^+ = f^-$, where $T_\#$ stands for the push-forward operator defined in Appendix B.2; equivalently, we can say that $T$ is a transport map if for all Borel sets $E \subseteq X$ one has

$$f^+\Big(\Big\{x \in X : T(x) \in E\Big\}\Big) = f^-(E) \,,$$

which is the exact analogue of (A.2) with $f^+ = \chi_A$ and $f^- = \chi_B$.

Another possible generalization is to consider a cost function $c : X \times X \to \mathbb{R}^+$ such that $c(x, y)$ represents the cost to move a unit mass from $x$ to $y$. Therefore, the cost associated with the transport map $T$ is simply

$$\int_X c(x, T(x))\, df^+(x)\,, \tag{A.3}$$

which again generalizes (A.1) with $c(x, y) = |y - x|$ when $X = \mathbb{R}^3$ and $f^+ = \chi_A$, $f^- = \chi_B$.

Despite the simplicity of the problem considered, it reveals to be very hard to attack. The main reasons are two: first of all, the set of the admissible maps has no good structure; indeed, even in the simplest case considered by Monge, the admissibility of $T$ relies on the validity of the highly non-linear equation

$$\big|\mathrm{Det}\, DT\big| = 1\,.$$

A second reason is that also the cost given by (A.1) or (A.3) depends in a quite involved way on the map $T$ so again there is no kind of linearity giving some help.

We can also note that, in general, the problem of finding a minimizer of the cost has no solution, and it can also easily happen that there are no admissible transport maps at all. For instance, if $f^+ = \delta_x$ is a Dirac mass at a point $x$, then for any map $T : X \to X$ one has $T_\# f^+ = \delta_{T(x)}$; therefore, there are no transport maps at all unless $f^-$ is a Dirac mass itself. In fact, the only obstacle to the existence of tranport maps is the presence of Dirac masses in $f^+$: indeed, the following theorem is well known (for a proof in the general case of Polish spaces one can refer to [59]).

**Theorem A.1.** *Let $X$ be a Polish space and $f^+$, $f^-$ be two positive Borel measures on $X$ with the same total mass. If $f^+$ is non-atomic (that is, for any $x \in X$ one has $f^+\big(\{x\}\big) = 0$) then there exist transport maps from $f^+$ to $f^-$.*

Even if the set of transport maps is nonempty, the existence of optimal transport maps may fail: consider for instance the case when $X = \mathbb{R}^2$, $f^+$ is the Hausdorff one-dimensional measure with density 1 on the segment $\{0\} \times [0, 1]$, and $f^-$ is the Hausdorff one-dimensional measure with density 1/2 on the two segments $\{\pm 1\} \times [0, 1]$; consider also the easiest cost function $c(x, y) = |y - x|$. In this case, it is immediate to understand that the infimum of the cost of the transport maps cannot be achieved, since otherwise an optimal transport map should split every point in the support of $f^+$ and move half of it in the left segment of the support of $f^-$ and half in the right one.

The first big step forward in the study of the mass transportation is due to the work by Kantorovich [42, 43]; to explain it, let us start from the consideration that, in both of the above examples, the obstacle was the impossibility to split the masses by a map. The idea of Kantorovich, then, was to allow this possibility; hence, the mass which is initially at $x$, should not be entirely transported to a point $T(x)$, rather it may be distributed on the support of $f^-$. Formally, this means that a new notion of transportation is introduced.

**Definition A.2.** A *transport plan* is a positive measure $\gamma \in \mathcal{M}^+(X\times X)$ such that the two projections $\pi_{1\#}\gamma$ and $\pi_{2\#}\gamma$ coincide with $f^+$ and $f^-$ respectively, where $\pi_1$ and $\pi_2$ are the projections of $X \times X$ on the first and second factor respectively.

It is important to notice the meaning of the above definition: the measure $\gamma$ corresponds to transporting a mass $\gamma(C\times D)$ from the set $C$ to the set $D$ for any choice of the sets $C$, $D \subseteq X$. Hence, every transport map $T$ correspond naturally to a transport plan $\gamma_T$, namely

$$\gamma_T = \big(\mathrm{Id}, T\big)_{\#} f^+ .$$

It is easy to generalize the cost of a transport map to transport plans: indeed, we will say that the cost of the transport plan $\gamma$ is

$$\iint_{X\times X} c(x,y)\, d\gamma(x,y) .$$

With this definition, the cost of the transport map $T$ equals the cost of the transport plan $\gamma_T$, so that the Kantorovich setting of the problem is indeed a generalization of the one by Monge.

It is immediate to see that both of the above mentioned difficulties for transport maps are immediately solved for transport plans: indeed, the transport plans form a bounded, convex and weakly* closed subset of the measures on $X \times X$, so they have a very good and perfectly known structure; in addition, the cost of the transport plans is linear with respect to $\gamma$.

Hence, in this new formulation, the existence of transports and of optimal transports always occurs. Indeed, the set of transport plans is nonempty, since the measure $f^+ \otimes f^-$ is always an admissible transport plan. Moreover, also the existence of optimal transport plans (i.e. those minimizing the cost) becomes very easy: in fact, if $\varphi$ is a lower semicontinuous function on the Polish space $Y$ and $\mu_n \stackrel{*}{\rightharpoonup} \mu$ are Borel measures on $Y$ then

$$\int_Y \varphi(y)\, d\mu(y) \leq \liminf \int_Y \varphi(y)\, d\mu_n(y) .$$

Therefore, provided that $c$ is lower semicontinuous, the problem of finding an optimal transport plan is trivial: it suffices to take any weak* limit of any minimizing sequence of transport plans.

Even if the Kantorovich formulation of the problem always admits optimal solutions, deducing the existence of transport maps remains a difficult issue. The main idea to show the existence of transport maps was found in the 70's by Sudakov [67]: indeed, in the case $c(x,y) = |y-x|$ with ambient space $X = \mathbb{R}^n$, it is known that any optimal transport plan moves the mass along non-intersecting segments. More precisely, if $\gamma$ is an optimal transport plan and $(x_1, y_1)$, $(x_2, y_2)$ belong to the support of $\gamma$, then the segments $x_1y_1$ and $x_2y_2$ cannot cross except in a common endpoint. Then $\mathbb{R}^n$ can be filled by non-intersecting segments (called *transport rays*) so that all the mass can be moved by the optimal transport plans only along those rays. Since it is possible to determine these segments knowing only $f^+$ and $f^-$, the idea of Sudakov was to consider the transport problem on each of these rays and then "glue" all the information found. This argument would reduce the mass transportation to one-dimensional problems, which are easily discussed, so this provides a possible strategy to show the existence of transport maps.

Nevertheless, two decades passed before the Sudakov argument were rigorously completed by means of fine tools of Geometric Measure Theory. One of the technical difficulties was the following: as we saw, the presence of Dirac masses can prevent the existence of optimal transport maps; however, even if $f^+$ is absolutely continuous with respect to the Lebesgue measure in $\mathbb{R}^n$, in principle it may happen that its "restrictions" to the transport rays have Dirac masses. By the way, also the definition of this restrictions needs to be carefully precised by means of measure disintegration. Anyway, Sudakov's formal idea has given a powerful strategy to attack the problem in the last years, and all the different proofs now available concerning the linear cost (i.e. $c(x, y) = |y - x|$) rely somehow on this idea.

The first proof of the existence of optimal transport maps was given independently by Brenier and Knott–Smith in the '80s [22, 23, 45], but instead of the Monge case of the linear cost, the quadratic one was considered.

**Theorem A.3.** *Consider the case $X = \mathbb{R}^n$ with the quadratic cost $c(x, y) = |y - x|^2$, and assume that $f^+(S) = 0$ for any set $S$ such that $\mathscr{H}^{n-1}(S) < +\infty$. Then there is a unique optimal transport plan, which in particular corresponds to an optimal transport map.*

In the following years this result has been widely generalized to other cases of strictly convex cost functions such as, for instance, $c(x, y) = |y - x|^p$ with $p > 1$; some references for these results are for instance [60, 40, 71].

The situation in the original Monge case is completely different, mainly due to the fact that the linear cost $c(x, y) = |y - x|$ is convex but not strictly convex. The first existence result for this case was given by Evans and Gangbo in 1999 in [36], and in the following few years their result was generalized in other papers [1, 70, 25]. We can summarize all their results in the following one.

**Theorem A.4.** *Assume that $X = \mathbb{R}^n$ and that $c(x, y) = |y - x|$, and let $f^+$ be a measure with compact support and absolutely continuous with respect to the Lebesgue measure. Then there exists an optimal transport map.*

We emphasize the big difference between the results for the linear and the quadratic costs. First of all, with the quadratic cost (or generally, with many strictly convex costs) one has the uniqueness of an optimal transport plan, which is also a map; on the other hand, with the linear cost there are many optimal transport plans, and many of them (but not all) are in particular transport maps. Moreover, while in the strictly convex case the essential assumption is that $f^+$ does not charge $(n-1)-$dimensional sets (in the sense of Theorem A.3), in the linear case it is essential that $f^+$ be absolutely continuous: the fact that this stronger assumption is indeed necessary is shown by a counterexample given in [4].

Also for the linear case many subsequent generalizations have been shown: the case where the ambient space $X$ is a manifold and $c(x,y) = d(x,y)$ is the distance on the manifold is considered in [39], while the case when $X = \mathbb{R}^n$ and $c(x,y) = \|y-x\|$ for a convex norm $\|\cdot\|$ different from the Euclidean one is considered in [25, 4, 3].

For the interested reader, there is a large number of books and wide surveys about the mass transportation problem from many different points of view. A very short list can be given by [71, 72, 60, 1, 4, 57].

# B Some Tools from Geometric Measure Theory

In this chapter we present some well known results about the definition of measures and their main properties and about some specific tools from Geometric Measure Theory, such as the Disintegration Theorem or the $\Gamma-$convergence; more precise and detailed results can be found in many complete books on these subjects, for instance in [33, 2].

## *B.1 Measures as Duals of the Continuous Functions*

We start from the following definition.

**Definition B.1.** A topological space $X$ is called a *Polish space if it is separable and metrizable with a distance making it complete.*

Recall that a space is said to be metrizable if it can endowed by some distance inducing the same topology; note also that the completeness, but not the separability, depends on the distance. In this paper, we always consider each Polish space endowed with its Borel $\sigma-$algebra, denoted by $\mathscr{B}(X)$.

Given a Polish space $X$, we consider the space $\mathcal{M}^+(X)$ of the finite positive Borel measures on it, which is defined as the set of all countably additive functions from $\mathscr{B}(X)$ to $\mathbb{R}^+$. It is well known that $\mathcal{M}^+(X)$ is a (strongly) closed subset of $\big(C_b(X)\big)'$, the dual space of the space $C_b(X)$ of the continuous and bounded real functions on $X$, endowed with the norm

$$\|u\|_{C_b(X)} := \sup\big\{|u(x)|,\ x \in X\big\}.$$

We denote by $\langle\cdot,\cdot\rangle$ the duality between $C_b(X)'$ and $C_b(X)$, and hence, in particular, the duality between $\mathcal{M}^+(X)$ and $C_b(X)$. Even though $\mathcal{M}^+(X)$ is not a linear space because we deal only with positive measures, we endow it with the norm induced by the inclusion in $C_b(X)'$, that is

$$\|\mu\|_{\mathcal{M}^+(X)} := \sup\left\{\langle\mu,u\rangle,\ \|u\|_{C_b(X)}=1\right\} = \mu(X).$$

The subspace of $C_b(X)'$ made by the increasing functionals, i.e. all those $\mu\in C_b(X)'$ such that $\langle\mu,u\rangle\geq 0$ whenever $u\geq 0$, is denoted by $C_b(X)'_+$: of course, the inclusion $\mathcal{M}^+(X)\subseteq C_b(X)'_+$ holds.

We can also consider on $\mathcal{M}^+(X)$ the weak* convergence induced by $C_b(X)$: we say that a sequence $\mu_n\in\mathcal{M}^+(X)$ weakly* converge to $\mu$, and we write $\mu_n \xrightarrow{*} \mu$, if

$$\langle\mu_n,u\rangle\longrightarrow\langle\mu,u\rangle \qquad \text{for each } u\in C_b(X).$$

**Definition B.2.** Given a positive measure $\eta\in\mathcal{M}^+(X)$, we define the set $\mathscr{B}_\eta(X)$ of the *measurable sets with respect to* $\eta$ to be the smallest $\sigma-$algebra containing both the set $\mathscr{B}(X)$ of all Borel sets and the set of all the $\eta$-negligible subsets of $X$. Moreover, we say that a set $B\subseteq X$ is *universally measurable* if $B\in\mathscr{B}_\eta(X)$ for any positive measure $\eta$.

**Definition B.3.** Given any positive measure $\mu\in\mathcal{M}^+(X)$ we define the *support of* $\mu$, written $\operatorname{spt}\mu$, to be the smallest closed set $K\subseteq X$ such that

$$\mu(X\setminus K)=0.$$

More in general, one says that the measure $\mu$ is *concentrated* on $\Gamma$ if

$$\mu(X\setminus\Gamma)=0.$$

For any measure $\mu$, the support $\operatorname{spt}\mu$ is a well-defined closed set of full measure; in particular, it is the smallest closed set of full measure; on the other hand, there are in general many sets on which $\mu$ is concentrated. However, it is often very useful to select some set of full measure for $\mu$ with particular properties, which may also differ from the support. In particular, a measure can be concentrated in a set much smaller than its support: if for example $X=\mathbb{R}$ and $\mu=\sum_i 2^{-i}\,\delta_{q_i}$ where $\mathbb{Q}=\{q_i\}$ is the set of the rationals, then $\mu$ is concentrated in the countable set $\mathbb{Q}$ but its support is the whole $\mathbb{R}$.

A first result on Polish spaces is given by the following

**Theorem B.4 (Ulam).** *Given a Polish space $X$ and $\mu\in\mathcal{M}^+(X)$, there is a $\sigma-$compact set in which $\mu$ is concentrated.*

This result is of primary importance, and can be found for example in [33]; one can refer to the same book for the proof of the following very important theorem.

**Theorem B.5 (Prokhorov).** *A bounded set $\{\mu_i, i \in I\}$ in $\mathcal{M}^+(X)$ is weakly* sequentially relatively compact if and only if for any $\varepsilon > 0$ there is a compact set $K_\varepsilon \subseteq X$ such that*

$$\mu_i(X \setminus K_\varepsilon) \leq \varepsilon \qquad \text{for any } i \in I. \tag{B.1}$$

The property (B.1) is called *tightness* of the set $\{\mu_i\}$; hence, Prokhorov Theorem says that a bounded set of measures is weakly* sequentially relatively compact if and only if the tightness property holds.

As we mentioned before, any measure is an element of $C_b(X)'_+$; however, it is fundamental to know when the opposite is also true: in fact, it is very often useful to define a measure $\mu$ by specifying the value of $\langle \mu, u \rangle$ for any $u \in C_b(X)$. When doing so, it is sufficient to check that $\langle \mu, \cdot \rangle$ is linear, continuous and increasing to derive that $\mu$ belongs to $C_b(X)'_+$, (and this usually follows trivially from the definition); but on the other hand, to establish that $\mu$ is in fact a measure is often not straightforward since the inclusion $\mathcal{M}^+(X) \subseteq C_b(X)'_+$ may in general be strict for Polish spaces.

Hence, we collect now some results which give necessary and sufficient conditions for an element of $C_b(X)'_+$ to be in $\mathcal{M}^+(X)$. We will notice that these results cover many important situations, in particular by Corollary B.8 they always apply to transport plans. The most important result to deal with the question whether an element of $C_b(X)'_+$ is a measure is given by Daniell's Theorem.

**Theorem B.6 (Daniell).** *Let $\mu \in C_b(X)'_+$: then $\mu \in \mathcal{M}^+(X)$ if and only if for every sequence $\{h_n\} \subseteq C_b(X)$ such that $h_n \searrow 0$ (i.e. $h_n$ is pointwise decreasing to 0) one has $\langle \mu, h_n \rangle \longrightarrow 0$.*

Note that, in the hypotheses above, $\langle \mu, h_n \rangle$ is a real positive decreasing sequence, then $\langle \mu, h_n \rangle \longrightarrow l$ for some $l \geq 0$; the Theorem claims that $\mu$ is a measure if and only if this limit $l$ is never strictly positive. Once again, one can find the proof of a more general assertion in the book [33]. We remark now some easy consequences of Daniell's Theorem.

**Proposition B.7.** *Let $\mu \in C_b(X)'_+$: then $\mu \in \mathcal{M}^+(X)$ if and only if for each $\varepsilon > 0$ there is a compact set $K$ such that $\langle \mu, u \rangle \leq \varepsilon \|u\|_{C_b(X)}$ whenever $u \in C_b(X)$, $u \geq 0$ on $X$, $u = 0$ on $K$.*

*Proof.* If $\mu \in \mathcal{M}^+(X)$ then the stated property immediately follows from Ulam's Theorem. On the other hand, assume that the property holds and take a sequence $h_n \searrow 0$ as in Daniell's Theorem: to show that $\mu \in \mathcal{M}^+(X)$ it is sufficient to check that $\langle \mu, h_n \rangle \to 0$. Fix then $\varepsilon > 0$ and consider the corresponding compact set $K$.

Since $K$ is compact, there is an integer $m$ such that $h_m \leq \varepsilon$ on $K$; let then $U$ be an open neighborhood of $K$ such that $h_m < 2\varepsilon$ on $U$. Notice that this is possible since $h_m$ is continuous, and that the open set $U$ depends on $\varepsilon$, on $K$ and on $m$. Take now a partition of the unity $\{\varphi_1, \varphi_2\}$ associated to the open sets $U$ and $X \setminus K$, i.e. $\varphi_i : X \to [0,1]$ is a continuous function for $i = 1, 2$, $\varphi_1 + \varphi_2 \equiv 1$, and $\varphi_1$ (resp. $\varphi_2$) is positive only inside the open set $U$ (resp. $X \setminus K$). We recall that this is always possible, since $X$ is metrizable: for example, if we denote by $d_1(x)$ and $d_2(x)$ the distance of $x$ from $X \setminus U$ and $K$ respectively, it suffices to note that $d_1$ and $d_2$ are continuous, that $d_1 + d_2$ is everywhere strictly positive and finally to define

$$\varphi_1(x) := \frac{d_1(x)}{d_1(x) + d_2(x)}, \qquad \varphi_2(x) := \frac{d_2(x)}{d_1(x) + d_2(x)}.$$

Now we write $h_m = g_1 + g_2$, where $g_i = \varphi_i h_m$: then we note that $g_1 \leq h_m$ on $U$ and $g_1 = 0$ outside of $U$: consequently $\|g_1\|_{C_b(X)} \leq 2\varepsilon$. On the other hand, $g_2 = 0$ on $K$ and, of course, $\|g_2\|_{C_b(X)} \leq \|h_m\| \leq \|h_1\|$, since the sequence $\{h_n\}$ is decreasing. Then, thanks to the hypothesis, we can evaluate

$$\langle \mu, h_m \rangle = \langle \mu, g_1 \rangle + \langle \mu, g_2 \rangle \leq 2\varepsilon \, \|\mu\|_{C_b(X)'_+} + \varepsilon \|h_1\|_{C_b(X)'_+} \leq \varepsilon \big(2\|\mu\| + \|h_1\|\big).$$

Since the sequence $\{h_n\}$ is decreasing and by the generality of $\varepsilon$ we deduce $\langle \mu, h_n \rangle \to 0$ and then the thesis follows by Daniell's Theorem. □

We can derive the following very useful corollary.

**Corollary B.8.** *Assume that $\gamma \in C_b(X \times Y)'_+$ and that $\pi_1\gamma \in \mathcal{M}^+(X)$ and $\pi_2\gamma \in \mathcal{M}^+(Y)$. Then $\gamma \in \mathcal{M}^+(X \times Y)$.*

*Proof.* By Proposition B.7, given an $\varepsilon > 0$ there are two compact sets $K_1 \subseteq X$ and $K_2 \subseteq Y$ such that $\langle \pi_1\gamma, v \rangle \leq \varepsilon \|v\|$ (resp. $\langle \pi_2\gamma, w \rangle \leq \varepsilon \|w\|$) whenever $v \in C_b(X)$, $v \geq 0$ on $X$, $v = 0$ on $K_1$ (resp. $w \in C_b(Y)$, $w \geq 0$ on $Y$, $w = 0$ on $K_2$). We consider then the compact set $K = K_1 \times K_2$ in $X \times Y$: given any $u \in C_b(X \times Y)$ such that $u \geq 0$ on $X$ and $u = 0$ on $K$, and given a $\delta > 0$, we can find an open neighborhood $U_1 \times U_2$ of $K$ such that $0 \leq u \leq \delta$ in $U_1 \times U_2$. Moreover, arguing as in Proposition B.7, we can find two functions $v : X \to [0, \|u\|]$ and $w : Y \to [0, \|u\|]$ such that $v = 0$ on $K_1$, $w = 0$ on $K_2$, $v = \|u\|$ out of $U_1$ and $w = \|u\|$ out of $U_2$. Since $v(x) + w(y) \geq u(x, y)$ whenever $(x, y) \notin U_1 \times U_2$ and $\langle \gamma, v(x) + w(y) \rangle = \langle \pi_1\gamma, v \rangle + \langle \pi_2\gamma, w \rangle$, arguing as in Proposition B.7 one derives

$$\langle \gamma, u \rangle \leq 2\,\varepsilon \|u\| + \delta \|\gamma\|;$$

by the generality of $\delta$, the thesis follows from Proposition B.7. □

The corollary above is very important in mass transportation problem: in fact, it implies in particular that any continuous and linear functional on $X \times Y$ with marginals $f^+$ and $f^-$ is in fact a measure, and hence it is a transport plan.

Let us now briefly discuss the problem whether or not the inclusion $\mathcal{M}^+(X) \subseteq C_b(X)'_+$ is strict: first of all, we have an immediate result.

**Lemma B.9.** *If $X$ is compact, then $\mathcal{M}^+(X) = C_b(X)'_+$.*

*Proof.* This trivially follows from Daniell's Theorem, since a sequence of continuous functions pointwise decreasing to 0 in a compact space is easily seen to be converging uniformly. □

However, the result above cannot be extended, as the following example shows.

*Example B.10.* Take $X = \mathbb{R}$, define $u_0 \equiv 1$ and for any integer $n \geq 1$ let $u_n : \mathbb{R} \to [0,1]$ be a continuous function such that $u_n \equiv 1$ on $[-n,n]$ and $u_n \equiv 0$ outside of $(-(n+1), n+1)$. Define $D$ to be the span of $\{u_n,\ n \geq 0\}$, that is a subspace of $C_b(X)$. Let moreover $\tilde{\mu}$ be the linear functional on $D$ given by $\langle \tilde{\mu}, u_0 \rangle = 1$ and $\langle \tilde{\mu}, u_n \rangle = 0$ for any $n \geq 1$, which is easily checked to be continuous and to have unit norm. By Hahn-Banach Theorem, there is an element $\mu$ of $C_b(X)'_+$ extending $\tilde{\mu}$. Since the functions $h_n = u_0 - u_n$ are as in the claim of Daniell's Theorem but $\langle \mu, h_n \rangle = 1$ for any $n$, $\mu$ is not a measure, then $\mu \in C_b(X)'_+ \setminus \mathcal{M}^+(X)$.

The example above can be clearly extended to cover all the non-compact Polish spaces, therefore in Lemma B.9 the opposite implication is valid too, so that $\mathcal{M}^+(X) = C_b(X)'_+$ if and only if $X$ is compact.

*Remark B.11.* Recall that Riesz Theorem claims that $\mathcal{M}^+(\mathbb{R})$ is exactly the set of all increasing linear functionals on $C_c(\mathbb{R})$. In general, using $C_c(X)$ in place of $C_b(X)$ to define a duality on $\mathcal{M}^+(X)$ has the advantage that Riesz Theorem, which states that $\mathcal{M}^+(X) = C_c(X)'_+$, is true in a wide generality: for example, making use of Daniell's Theorem, one can quite easily show that this is true whenever $X$ is a countable union of open sets $\{U_n\}$ with compact closures $\{K_n\}$ such that $K_n \subseteq U_{n+1}$ for any integer $n$; in particular, Riesz Theorem is true whenever $X$ is a locally compact Polish space. Notice in particular that, since $C_c(X) \subseteq C_b(X)$, whenever Riesz Theorem holds for $X$ one has that for any $\mu \in C_b(X)'_+$ there is a $\tilde{\mu} \in \mathcal{M}^+(X)$ for which one has $\langle \mu, u \rangle = \langle \tilde{\mu}, u \rangle$ for every $u \in C_c(X)$.

However, it is often preferable to use the duality with $C_b(X)$ whenever one wants to deal with non compactly supported measures. In this book, the two choices are equivalent since we assume our ambient space to be a compact subset of $\mathbb{R}^n$.

We conclude this section by stating two other important results about Polish spaces, which can be found for example in [28] (in a more general version).

**Theorem B.12 (Measurable selection).** *Let $X$ and $Y$ be two Polish spaces, and let $\eta$ be a Borel measure on $X$. If $\Delta \subseteq X \times Y$ is $\mathscr{B}_\eta(X) \otimes \mathscr{B}(Y)$-measurable and the projection of $\Delta$ on $X$ is a set of full $\eta-$measure, then there exists a measurable selection of $\Delta$, i.e. an $\eta-$measurable function $\sigma : X \to Y$ such that for $\eta-$a.e. $x \in X$ one has $(x, \sigma(x)) \in \Delta$.*

**Theorem B.13 (Projection).** *Let $X$ and $Y$ be two Polish spaces, and let $\Delta \subseteq X \times Y$ be a Borel set; then the projection of $\Delta$ on $X$ is universally measurable.*

## *B.2 Push-forward and Tensor Product of Measures*

Given a measurable function $\varphi : X \to Y$, we may define a linear and mass-preserving operator $\varphi_\# : \mathcal{M}^+(X) \to \mathcal{M}^+(Y)$ (by mass-preserving we mean that $\|\mu\|_{\mathcal{M}^+(X)} = \|\varphi_\#\mu\|_{\mathcal{M}^+(Y)}$), according to the following formula:

$$\varphi_\#\mu(B) := \mu\big(\varphi^{-1}(B)\big) \qquad \forall\, A \in \mathscr{B}\,(Y).$$

It is easily noticed that $\varphi_\#\mu \in \mathcal{M}^+(Y)$, and in particular for any $\alpha \in C_b(Y)$

$$\langle \varphi_\#\mu, \alpha\rangle = \int_Y \alpha(y)\, d\varphi_\#\mu(y) = \int_X \alpha\big(\varphi(x)\big)\, d\mu(x).$$

An immediate but very useful consequence of the definition is that for any $\varphi : X \to Y$ and $\psi : Y \to Z$ it is

$$(\psi \circ \varphi)_\#\mu = \psi_\#(\varphi_\#\mu). \tag{B.2}$$

Given two measures $\mu \in \mathcal{M}^+(X)$ and $\nu \in \mathcal{M}^+(Y)$, we define the *tensor product* of $\mu$ and $\nu$, $\mu \otimes \nu \in \mathcal{M}^+(X \times Y)$, as the unique measure on $X \times Y$ satisfying

$$\mu \otimes \nu(A \times B) := \mu(A) \cdot \nu(B) \qquad \forall\, A \in \mathscr{B}\,(X),\ B \in \mathscr{B}\,(Y).$$

The fact that there is a measure satisfying the above property follows from Fubini-Tonelli Theorem, while the uniqueness occurs because the smallest $\sigma-$algebra containing all the sets $A \times B$ with $A \in \mathscr{B}\,(X)$, $B \in \mathscr{B}\,(Y)$ is the whole $\mathscr{B}\,(X \times Y)$. In particular, the marginals of $\mu \otimes \nu$ are

$$\pi_{1\#}(\mu \otimes \nu) = \nu(Y)\mu, \qquad\qquad \pi_{2\#}(\nu \otimes \mu) = \mu(X)\nu\,.$$

An interesting particular case is when $\mu$ and $\nu$ are probability measures; so, the projections of $\mu \otimes \nu$ are precisely $\mu$ and $\nu$: this is particularly useful in the study of the mass transportation, since if $\|f^+\| = \|f^-\| = 1$ then $f^+ \otimes f^-$ is always a transport plan between $f^+$ and $f^-$.

## *B.3 Measure Valued Maps and Disintegration Theorem*

Here we briefly introduce a couple of tools concerning the measure-valued maps and then present the Disintegration Theorem.

**Definition B.14.** A map $\tau : X \to \mathcal{M}^+(Y)$ is called Borel measure-valued map (resp. $\mu-$measurable measure valued map, where $\mu \in \mathcal{M}^+(X)$) if for any Borel set $B \subseteq Y$ the function $x \to \tau(x)(B)$ is Borel (resp. $\mu-$measurable). Equivalently, $\tau$ is said to be Borel (resp. $\mu-$measurable) if for any bounded Borel $\varphi : X \times Y \to \mathbb{R}$ the function

$$x \to \int_Y \varphi(x,y)\, d\tau(x)(y)$$

is Borel (resp. $\mu-$measurable).

Thanks to the notion above, we can generalize the idea of a tensor product between two measures: consider a measure $\nu \in \mathcal{M}^+(Y)$ and a $\nu-$measurable measure valued map $y \mapsto \gamma_y$ with $\gamma_y \in \mathcal{M}^+(X)$. We define the *tensor product* between $\{\gamma_y\}$ and $\nu$ to be the measure $\gamma \in \mathcal{M}^+(X)$ given by the formula

$$\langle \gamma, \varphi \rangle := \int_Y \langle \gamma_y, \varphi \rangle\, d\nu(y)$$

for any $\varphi \in C_b(X)$. We always denote this measure by

$$\gamma := \gamma_x \otimes \nu.$$

We can now present the Disintegration Theorem: this result allows to decompose a measure $\gamma$ over the space $X$ with respect to a Borel function $\alpha : X \to Y$, where this "decomposition" is intended as a tensor product between suitable probability measures and the push-forward of $\gamma$. The proof of this Theorem can be found, for instance, in [2] or [33].

**Theorem B.15 (Disintegration).** *Let $\alpha : X \to Y$ be a given Borel map and $\gamma \in \mathcal{M}^+(X)$ is a given measure, and define $\mu \in \mathcal{M}^+(Y)$ by setting $\mu := \alpha_{\#}\gamma$. Then there exists a $\mu-$measurable measure valued function $y \mapsto \gamma_y$ such that $\gamma_y$ is a probability measure on $X$ for any $y$ and*

*(i)* $\gamma = \gamma_y \otimes \mu$;
*(ii)* $\gamma_y$ *is concentrated on* $\{x : \alpha(x) = y\}$ *for* $\mu-$*a.e.* $y \in Y$.

*Moreover, the measures $\gamma_y$ are uniquely determined by (i) and (ii) for $\mu-$a.e. $y \in Y$.*

We state and prove now a useful consequence of the above theorem.

**Lemma B.16.** *The operations of disintegration and of composition commute, i.e. if*

$$\gamma = \gamma_y \otimes \alpha_{\#}\gamma$$

*is the disintegration of $\gamma \in \mathcal{M}^+(X)$ with respect to some $\alpha : X \to Y$ and a function $\beta : X \to Z$ is given, then*

$$\beta_{\#}\gamma = \beta_{\#}\gamma_y \otimes \alpha_{\#}\gamma. \tag{B.3}$$

*In particular, if* $\alpha = \delta \circ \beta$ *for some* $\delta : Z \to Y$ *and*

$$\beta_{\#}\gamma = \mu_y \otimes \delta_{\#}(\beta_{\#}\gamma) \tag{B.4}$$

*is the disintegration of* $\beta_{\#}\gamma$ *with respect of* $\delta$*, for a.e.* $y \in Y$ *it is*

$$\mu_y = \beta_{\#}\gamma_y. \tag{B.5}$$

*Proof.* The first part is easy: for any $\varphi \in C_c(Z)$, recalling the properties of the push-forward one has

$$\begin{aligned}\langle \beta_{\#}\gamma, \varphi\rangle &= \int_X \varphi(\beta(x))\, d\gamma(x) = \int_Y \Big(\int_X \varphi(\beta(x))\, d\gamma_y(x)\Big)\, d\nu(y)\\ &= \int_Y \Big(\int_Z \varphi(z)\, d\beta_{\#}\gamma_y(z)\Big)\, d\nu(y) = \langle \beta_{\#}\gamma_y \otimes \nu, \varphi\rangle,\end{aligned}$$

thus the claim follows.

Concerning the second part, by the properties of disintegration (B.3) becomes

$$\beta_{\#}\gamma = \beta_{\#}\gamma_y \otimes \big(\delta_{\#}\beta_{\#}\gamma\big).$$

Recall now the disintegration (B.4): according to Theorem B.15, for a.e. $y \in Y$ the measure $\mu_y$ is concentrated on the set $\{z \in Z : \delta(z) = y\}$; analogously, $\gamma_y$ is concentrated on the set $\{x \in X : \alpha(x) = y\} = \{x \in X : \delta(\beta(x)) = y\}$. But then $\beta_{\#}\gamma_y$ is concentrated on

$$\beta\Big(\big\{x \in X :\ \delta(\beta(x)) = y\big\}\Big) \subseteq \big\{z \in Z :\ \delta(z) = y\big\};$$

moreover, $\alpha_{\#}\gamma = \delta_{\#}(\beta_{\#}\gamma)$ by (B.2). Then, by the uniqueness part of Theorem B.15, we infer the validity of (B.5) and hence also the second claim is achieved. □

## B.4 $\Gamma-$convergence

In this section we briefly recall the definition and the main properties of the $\Gamma-$convergence; for a more complete and precise reference we address the reader to the books [29, 12].

The notion of $\Gamma-$convergence, first proposed by De Giorgi in [31, 32], is the following: let $X$ be a metric space, and assume that we are given a sequence of functionals $g_n : X \to \overline{\mathbb{R}}$ and a functional $g : X \to \overline{\mathbb{R}}$. We say that $g_n$ $\Gamma-$converges to $g$, or $g_n \xrightarrow{\Gamma} g$, if the following hold:

$$\textbf{(i)} \qquad \forall\, x,\ \forall\, \{x_n\} \to x, \quad g(x) \le \liminf_{n\to\infty} g_n(x_n)\,;$$

$$\textbf{(ii)} \qquad \forall\, x,\ \exists\, \{x_n\} \to x : \quad g(x) \ge \limsup_{n\to\infty} g_n(x_n)\,.$$

The first property is usually called the *liminf inequality* and the second one *limsup inequality*. Note that thanks to **(i)**, one could simply write in **(ii)** $g(x) = \lim g_n(x_n)$ instead of $g(x) \geq \limsup g_n(x_n)$. Moreover, given an $x \in X$, any sequence $\{x_n\} \to x$ for which the property **(ii)** is fulfilled is called *recovery sequence*. The first fundamental property that one can immediately notice is the following.

**Proposition B.17.** *If $g_n \xrightarrow{\Gamma} g$ and there exists a compact set $K \subseteq X$ so that for any $n \in \mathbb{N}$ one has $\inf_X g_n = \inf_K g_n$, then $g$ admits a minimum and $\inf g_n \to \min g$. Moreover, for any sequence $x_n$ such that $g_n(x_n) - \inf g_n \to 0$ and that $x_n \to \bar{x}$, one has that $\bar{x}$ is a minimum point for $g$.*

*Proof.* It suffices to take $x_n \in K$ so that $g_n(x_n) \leq \inf g_n + 1/n$; by the compactness of $K$ we know the existence of a subsequence $\{n_i\}_{i\in\mathbb{N}}$ such that $x_{n_i} \to \bar{x}$ for a certain $\bar{x} \in K$: moreover, we can also assume that

$$g_{n_i}(x_{n_i}) \xrightarrow[i\to\infty]{} \liminf_{n\to\infty} \inf_X g_n \,.$$

By the liminf property we know then that

$$g(\bar{x}) \leq \liminf_{i\to\infty} g_{n_i}(x_{n_i}) = \liminf_{n\to\infty} \inf_X g_n \,. \tag{B.6}$$

On the other hand, take any $\tilde{x} \in X$: by the limsup property we know the existence of a sequence $x_n \in X$ for which $x_n \to \tilde{x}$ and

$$g(\tilde{x}) \geq \limsup_{n\to\infty} g_n(x_n) \geq \limsup_{n\to\infty} \inf_X g_n \,. \tag{B.7}$$

From (B.6) and (B.7) we deduce that $\bar{x}$ is a minimum point for $g$, as well as that $\inf_X g_n$ converges, for $n \to \infty$, to $\min_X g$. The thesis then immediately follows. □

We claim now the second property, which is also very important, namely a compactness result for $\Gamma$-convergence.

**Theorem B.18.** *Assume that $X$ is separable. Then, for any sequence of functions $g_n : X \to \overline{\mathbb{R}}$, there exists a subsequence $g_{n_i}$ which admits a $\Gamma$-limit.*

The above result is very strong: indeed, given any sequence of functions, it allows us to assume, up to a subsequence, that they $\Gamma$-converge to some limit.

We present now the definition of the $\Gamma - \liminf$ and $\Gamma - \limsup$ of a sequence of functions. Given a sequence $\{g_n\}$, we define

$$\begin{aligned} \Gamma - \liminf_{n\to\infty} g_n(x) &:= \inf \Bigl\{ \liminf_{n\to\infty} g_n(x_n) : \, x_n \to x \Bigr\} ; \\ \Gamma - \limsup_{n\to\infty} g_n(x) &:= \inf \Bigl\{ \limsup_{n\to\infty} g_n(x_n) : \, x_n \to x \Bigr\} . \end{aligned} \tag{B.8}$$

It is clear from the definitions that one has always

$$\Gamma - \liminf g_n \leq \Gamma - \limsup g_n \,,$$

and that the sequence $\Gamma-$converges (to a function $g$) if and only if

$$\Gamma - \liminf g_n = \Gamma - \limsup g_n \quad (= g)\,.$$

**Proposition B.19.** *One has*

$$\Gamma - \liminf_{n\to\infty} g_n = \sup_{n\in\mathbb{N}} \Big(\mathrm{env}\big(\inf_{m\geq n} g_m\big)\Big)\,, \tag{B.9}$$

*where* $\mathrm{env}(\varphi)$ *denotes the lower semicontinuous envelope of any function* $\varphi : X \to \overline{\mathbb{R}}$. *In particular,* $\Gamma - \liminf g_n$ *is always lower semicontinuous in* $X$.

*Proof.* The equivalence (B.9) is verified directly from the definition (B.8). The lower semicontinuity of the $\Gamma - \liminf$ follows from (B.9) once one reminds that the supremum of lower semicontinuous functions is still lower semicontinuous. □

Finally, one can show an important property of the $\Gamma - \liminf$ of the sequence $g_n$: it is the infimum of all the possible $\Gamma-$limits of subsequences of $\{g_n\}$; analagously, the $\Gamma - \limsup$ is the supremum of all the possible $\Gamma-$limits of subsequences of $\{g_n\}$.

We conclude this section pointing out an useful consequence of (B.9) in the setting of the weak* convergence of measures.

**Lemma B.20.** *Let* $X$ *be a Polish space,* $\{\nu_n\} \in \mathcal{M}^+(X)$ *a sequence of measures weakly* converging to* $\nu$, *and* $\{g_n\} : X \to \mathbb{R}$ *a sequence of l.s.c. functions. Then*

$$\int_X \Gamma - \liminf_{n\to\infty} g_n \, d\nu \leq \liminf_{n\to\infty} \int_X g_n \, d\nu_n \,.$$

*Proof.* Defining for simplicity

$$\tau_n := \mathrm{env}\big(\inf_{m\geq n} g_m\big)\,,$$

we fix $j \in \mathbb{N}$ and evaluate

$$\begin{aligned}
\liminf_{n\to\infty} \int_X g_n \, d\nu_n &\geq \liminf_{n\to\infty} \int_X \big(\inf_{m\geq j} g_m\big)\, d\nu_n \\
&\geq \liminf_{n\to\infty} \int_X \mathrm{env}\big(\inf_{m\geq j} g_m\big)\, d\nu_n \\
&= \liminf_{n\to\infty} \int_X \tau_j \, d\nu_n \geq \int_X \tau_j \, d\nu.
\end{aligned}$$

Since this is true for any $j \in \mathbb{N}$, by the Lebesgue monotone convergence theorem and (B.9) the thesis follows. □

# References

1. L. Ambrosio, Lecture Notes on Optimal Transport Problems, in "Mathematical Aspects of Evolving Interfaces", Lecture Notes in Mathematics, LNM **1812**, Springer (2003), 1–52.
2. L. Ambrosio, N. Fusco, & D. Pallara, Functions of Bounded Variation and Free Discontinuity Problems, Oxford mathematical monographs, Oxford University Press, Oxford, 2000.
3. L. Ambrosio, B. Kirchheim & A. Pratelli, Existence of optimal transport maps for crystalline norms, Duke Math. J. **125** (2004), no. 2, 207–241.
4. L. Ambrosio & A. Pratelli, Existence and stability results in the $L^1$ theory of optimal transportation, in "Optimal Transportation and Applications", Lecture Notes in Mathematics, LNM **1813**, Springer (2003), 123–160.
5. L. Ambrosio & P. Tilli, Topics on analysis in metric spaces, Oxford Lecture Series in Mathematics and its Applications, vol. **25**, Oxford University Press, Oxford (2004).
6. V. Bangert, Minimal measures and minimizing closed normal one-currents, Geom. Funct. Anal., **9** (1999), 413–427.
7. M. Beckmann, T. Puu, Spatial economics: density, potential and flow, Elsevier Science Ltd., 1985.
8. M. Bernot, V. Caselles, & J.-M. Morel, Traffic plans, Publ. Mat., **49** (2) (2005), 417–451.
9. M. Bernot, V. Caselles & J.-M. Morel, Are there infinite irrigation trees?, J. Mathematical Fluid Mechanics, **8** (3) (2006), 311–332.
10. M. Bernot, V. Caselles & J.-M. Morel., Optimal transportation network, Lecture Notes in Mathematics, **1955**, Springer, 2008.
11. S. Bhaskaran & F.J.M. Salzborn, Optimal design of gas pipeline networks, J. Oper. Res. Society **30** (1979), 1047–1060.
12. A. Braides, $\Gamma$-convergence for beginners, Oxford Lecture Series in Mathematics and its Applications, **22**, Oxford University Press, 2002.
13. A. Brancolini, Problemi di Ottimizzazione in Teoria del Trasporto e Applicazioni, degree thesis, Università di Pisa, Pisa (2002). Available at the web page http://cvgmt.sns.it.
14. A. Brancolini & G. Buttazzo, Optimal networks for mass transportation problems, ESAIM Control Optim. Calc. Var., **11** (2005), 88–101.
15. A. Brancolini, G. Buttazzo & F. Santambrogio, Path functionals over Wasserstein spaces, J. Eur. Math. Soc. **8** (2006), 415–434.
16. G. Buttazzo, Semicontinuity, Relaxation and Integral Representation in the Calculus of Variations, Pitman Res. Notes Math. Ser. **207**, Longman, Harlow (1989).
17. G. Buttazzo, E. Oudet & E. Stepanov, Optimal transportation problems with free Dirichlet regions, in "Variational Methods for Discontinuous Structures", Cernobbio

2001, Progress in Nonlinear Differential Equations **51**, Birkhäuser Verlag, Basel (2002), 41–65.
18. G. Buttazzo, A. Pratelli & E. Stepanov, Optimal pricing policies for public transportation networks, SIAM J. Optimization, **16** (3) (2006), 826–853.
19. G. Buttazzo & E. Stepanov, On regularity of transport density in the Monge-Kantorovich problem, SIAM J. Control Optim., **42** (3) (2003), 1044–1055.
20. G. Buttazzo & E. Stepanov, Minimization problems for average distance functionals, in "Calculus of Variations: topics from the mathematical heritage of E. De Giorgi", Quaderni di Matematica II Università di Napoli, vol. **14**, Aracne Editrice, Roma (2004), 47–84.
21. G. Buttazzo & E. Stepanov, Optimal transportation networks as free Dirichlet regions for the Monge-Kantorovich problem, Ann. Scuola Norm. Sup. Pisa Cl. Sci., (5) **2** (2003), 631–678.
22. Y. Brenier, Décomposition polaire et réarrangement monotone des champs de vecteurs, C. R. Acad. Sci. Paris Sér. I Math., **305** no. 19 (1987), 805–808.
23. Y. Brenier, Polar factorization and monotone rearrangement of vector-valued functions, Comm. Pure Appl. Math., **44** no. 4 (1991), 375–417.
24. Y. Brenier, The dual least action problem for an ideal, incompressible fluid, Arch. Rational Mech. Anal., **122** (4) (1993), 323–351.
25. L. Caffarelli, M. Feldman & R.J. McCann, Constructing optimal maps for Monge's transport problem as a limit of strictly convex costs, J. Amer. Math. Soc., **15** (2002), 1–26.
26. P. Cannarsa & P. Cardaliaguet, Representation of equilibrium solutions to the table problem for growing sandpiles, J. Eur. Math. Soc. (JEMS), **6** (4) (2004), 435–464.
27. V. Caselles & J.-M. Morel, Irrigation, in Variational methods for discontinuous structures, Vol. **51** of *Progr. Nonlinear Differential Equations Appl.*, pages 81–90, Birkhäuser, Basel, 2002.
28. C. Castaing & M. Valadier, Convex analysis and measurable multifunctions, Lecture Notes in Mathematics, **580**, Springer (1977).
29. G. Dal Maso, An introduction to $\Gamma$-convergence, Progress in Nonlinear Differential Equations and their Applications, Birkhäuser, 1993.
30. G. Dal Maso & R. Toader, A Model for the quasi-static growth of brittle fractures: existence and approximation results, Arch. Rational Mech. Anal., **162** (2002), 101–135.
31. E. De Giorgi, Sulla convergenza di alcune successioni d'integrali del tipo dell'area (Italian), Rend. Mat. **8** (1975), 277–294.
32. E. De Giorgi, T. Franzoni, Su un tipo di convergenza variazionale (Italian), Atti Accad. Naz. Lincei Rend. Cl. Sci. Fis. Mat. Natur. **58** (1975), no. 6, 842–850.
33. C. Dellacherie & P.A. Meyer, Probabilities and potential, North-Holland Publ. Co. (1978).
34. G. Devillanova & S. Solimini, On the dimension of an irrigable measure, Rend. Sem. Mat. Univ. Padova, **117** (2007), 1–49.
35. L.C. Evans, Partial Differential Equations and Monge–Kantorovich Mass Transfer, Current Developments in Mathematics (1997), 65–126.
36. L.C. Evans & W. Gangbo, Differential Equations Methods for the Monge–Kantorovich Mass Transfer Problem, Memoirs of the A.M.S., Vol. **137**, Number 653, (1999).
37. K.J. Falconer, The geometry of fractal sets, Cambridge University Press (1985).
38. M. Feldman & R.J. McCann, Uniqueness and transport density in Monge's mass transportation problem, Calc. Var. P.D.E. **15** (2002), 81–113.
39. M. Feldman & R.J. McCann, Monge's transport problem on a Riemannian manifold, Trans. Amer. Math. Soc. **354** (2002), 1667–1697.
40. W. Gangbo & R.J. McCann, The geometry of optimal transportation, Acta Math., **177** (1996), 113–161.
41. E.N. Gilbert, Minimum cost communication networks, Bell System Tech. J. **46** (1967), 2209–2227.

42. L.V. Kantorovich, On mass transportation, Doklady Acad. Sci. USSR. **37** (7–8) (1942), 227–229 (in Russian).
43. L.V. Kantorovich, On a problem of Monge, Uspekhi Mat. Nauk. **3** (1948), 225–226 (in Russian).
44. L.V. Kantorovich, Economic models of best use of resources, Acad. Sci USSR, 1960 (in Russian).
45. M. Knott & C. Smith, On the optimal mapping of distributions, J. Optim. Theory Appl., **43** no. 1 (1984), 39–49.
46. C. Kuratowski, Topologie, vol. **1**, Państwowe Wydawnictwo Naukowe, Warszawa (1958).
47. D.H. Lee, Low cost drainage networks, Networks **6** (1976), 351–371.
48. F. Maddalena, J.M. Morel & S. Solimini, A variational model of irrigation patterns, Interfaces Free Bound **5** (4) (2003), 391–415.
49. G. Monge, Memoire sur la Theorie des Déblais et des Remblais, Hist. de l'Acad. des Sciences de Paris (1781).
50. J.-M. Morel & F. Santambrogio, Comparison of distances between measures, Appl. Math. Lett. **20** (2007), no. 4, 427–432.
51. F. Morgan & R. Bolton, Hexagonal economic regions solve the location problem, Amer. Math. Monthly **109** (2) (2002), 165–172.
52. S.J.N. Mosconi & P. Tilli, $\Gamma$-convergence for the irrigation problem, J. Convex Anal. **12** (1) (2005), 145–158.
53. J.C. Oxtoby, Homeomorphic measures in metric spaces, Proc. Amer. Math. Soc., **24** (1970), 419–423.
54. E. Paolini & E. Stepanov, Qualitative properties of maximum distance minimizers and average distance minimizers in $\mathbf{R}^n$, J. Math. Sciences (N.Y.), **122** (3) (2004), 105–122.
55. E. Paolini & E. Stepanov, Optimal transportation networks as flat chains, Interfaces Free Bound., **8** (2006) 393–436.
56. E. Paolini & E. Stepanov, Connecting measures by means of branched transportation networks at finite cost, preprint (2006).
57. A. Pratelli, Existence of optimal transport maps and regularity of the transport density in mass transportation problems, Ph.D. Thesis, Scuola Normale Superiore, Pisa, Italy (2003). Avalaible on http://cvgmt.sns.it/ .
58. A. Pratelli, Equivalence between some definitions for the optimal mass transport problem and for the transport density on manifolds, Ann. Mat. Pura Appl., **184** (2005), no. 2, 215–238.
59. A. Pratelli, On the equality between Monge's infimum and Kantorovich's minimum in optimal mass transportation, Ann. Inst. H. Poincaré Probab. Statist., **43** (2007), no. 1, 1–13.
60. S.T. Rachev & L. Rüschendorf, Mass Transportation Problems, Springer–Verlag (1998).
61. H.L. Royden, Real Analysis (2nd edition), Macmillan, (1968).
62. F. Santambrogio, Optimal channel networks, landscape function and branched transport, Interfaces Free Bound., **9** (2007), no. 1, 149–169.
63. F. Santambrogio & P. Tilli, Blow-up of optimal sets in the irrigation problem, Journal of Geometric Analysis, **15** (3) (2005), 343–362.
64. L. Simon, Lectures on Geometric Measure Theory, Proceedings of the Centre for Mathematical Analysis, Australian National University, Volume **3** (1983).
65. S.K. Smirnov, Decomposition of solenoidal vector charges into elementary solenoids and the structure of normal one-dimensional currents, St. Petersburg Math. J., **5** (4) (1994), 841–867.
66. E. Stepanov, Partial geometric regularity of some optimal connected transportation networks, J. Math. Sciences (N.Y.), **132** (4) (2006), 522–552.
67. V.N. Sudakov, Geometric Problems in the Theory of Infinite–Dimensional Probability Distributions, Proc. of the Steklov Institute of Mathematics, **141** (1979).
68. A. Suzuki & Z. Drezner, The $p$-center location, Location science **4** (1–2) (1996), 69–82.

69. A. Suzuki & A. Okabe, Using Voronoi diagrams, in *Facility location: a survey of applications and methods*, Z. Drezner editor, Springer series in operations research, pages 103–118. Springer Verlag, 1995.
70. N.S. Trudinger & X.J. Wang, On the Monge mass transfer problem, Calc. Var. P.D.E., **13** (2001), 19–31.
71. C. Villani, Topics in mass transportation, Graduate Studies in Mathematics, 58. American Mathematical Society, Providence, RI, 2003. xvi+370 pp.
72. C. Villani, Optimal transport, old and new, Grundlehren dem Mathematischen Wissenshaften, Springer (to appear).
73. Q. Xia, Optimal paths related to transport problems, Communications in Contemporary Math., **5** (2) (2003), 251–279.
74. Q. Xia, Interior regularity of optimal transport paths, Calc. Var. Partial Diff. Equations, **20** (3) (2004), 283–299.

# Index

# Lecture Notes in Mathematics

For information about earlier volumes
please contact your bookseller or Springer
LNM Online archive: springerlink.com

Vol. 1778: H. Kiechle, Theory of K-Loops (2002)
Vol. 1779: I. Chueshov, Monotone Random Systems (2002)
Vol. 1780: J. H. Bruinier, Borcherds Products on O(2,1) and Chern Classes of Heegner Divisors (2002)
Vol. 1781: E. Bolthausen, E. Perkins, A. van der Vaart, Lectures on Probability Theory and Statistics. Ecole d' Eté de Probabilités de Saint-Flour XXIX-1999. Editor: P. Bernard (2002)
Vol. 1782: C.-H. Chu, A. T.-M. Lau, Harmonic Functions on Groups and Fourier Algebras (2002)
Vol. 1783: L. Grüne, Asymptotic Behavior of Dynamical and Control Systems under Perturbation and Discretization (2002)
Vol. 1784: L. H. Eliasson, S. B. Kuksin, S. Marmi, J.-C. Yoccoz, Dynamical Systems and Small Divisors. Cetraro, Italy 1998. Editors: S. Marmi, J.-C. Yoccoz (2002)
Vol. 1785: J. Arias de Reyna, Pointwise Convergence of Fourier Series (2002)
Vol. 1786: S. D. Cutkosky, Monomialization of Morphisms from 3-Folds to Surfaces (2002)
Vol. 1787: S. Caenepeel, G. Militaru, S. Zhu, Frobenius and Separable Functors for Generalized Module Categories and Nonlinear Equations (2002)
Vol. 1788: A. Vasil'ev, Moduli of Families of Curves for Conformal and Quasiconformal Mappings (2002)
Vol. 1789: Y. Sommerhäuser, Yetter-Drinfel'd Hopf algebras over groups of prime order (2002)
Vol. 1790: X. Zhan, Matrix Inequalities (2002)
Vol. 1791: M. Knebusch, D. Zhang, Manis Valuations and Prüfer Extensions I: A new Chapter in Commutative Algebra (2002)
Vol. 1792: D. D. Ang, R. Gorenflo, V. K. Le, D. D. Trong, Moment Theory and Some Inverse Problems in Potential Theory and Heat Conduction (2002)
Vol. 1793: J. Cortés Monforte, Geometric, Control and Numerical Aspects of Nonholonomic Systems (2002)
Vol. 1794: N. Pytheas Fogg, Substitution in Dynamics, Arithmetics and Combinatorics. Editors: V. Berthé, S. Ferenczi, C. Mauduit, A. Siegel (2002)
Vol. 1795: H. Li, Filtered-Graded Transfer in Using Noncommutative Gröbner Bases (2002)
Vol. 1796: J.M. Melenk, hp-Finite Element Methods for Singular Perturbations (2002)
Vol. 1797: B. Schmidt, Characters and Cyclotomic Fields in Finite Geometry (2002)
Vol. 1798: W.M. Oliva, Geometric Mechanics (2002)
Vol. 1799: H. Pajot, Analytic Capacity, Rectifiability, Menger Curvature and the Cauchy Integral (2002)
Vol. 1800: O. Gabber, L. Ramero, Almost Ring Theory (2003)
Vol. 1801: J. Azéma, M. Émery, M. Ledoux, M. Yor (Eds.), Séminaire de Probabilités XXXVI (2003)
Vol. 1802: V. Capasso, E. Merzbach, B. G. Ivanoff, M. Dozzi, R. Dalang, T. Mountford, Topics in Spatial Stochastic Processes. Martina Franca, Italy 2001. Editor: E. Merzbach (2003)
Vol. 1803: G. Dolzmann, Variational Methods for Crystalline Microstructure – Analysis and Computation (2003)
Vol. 1804: I. Cherednik, Ya. Markov, R. Howe, G. Lusztig, Iwahori-Hecke Algebras and their Representation Theory. Martina Franca, Italy 1999. Editors: V. Baldoni, D. Barbasch (2003)
Vol. 1805: F. Cao, Geometric Curve Evolution and Image Processing (2003)
Vol. 1806: H. Broer, I. Hoveijn. G. Lunther, G. Vegter, Bifurcations in Hamiltonian Systems. Computing Singularities by Gröbner Bases (2003)
Vol. 1807: V. D. Milman, G. Schechtman (Eds.), Geometric Aspects of Functional Analysis. Israel Seminar 2000-2002 (2003)
Vol. 1808: W. Schindler, Measures with Symmetry Properties (2003)
Vol. 1809: O. Steinbach, Stability Estimates for Hybrid Coupled Domain Decomposition Methods (2003)
Vol. 1810: J. Wengenroth, Derived Functors in Functional Analysis (2003)
Vol. 1811: J. Stevens, Deformations of Singularities (2003)
Vol. 1812: L. Ambrosio, K. Deckelnick, G. Dziuk, M. Mimura, V. A. Solonnikov, H. M. Soner, Mathematical Aspects of Evolving Interfaces. Madeira, Funchal, Portugal 2000. Editors: P. Colli, J. F. Rodrigues (2003)
Vol. 1813: L. Ambrosio, L. A. Caffarelli, Y. Brenier, G. Buttazzo, C. Villani, Optimal Transportation and its Applications. Martina Franca, Italy 2001. Editors: L. A. Caffarelli, S. Salsa (2003)
Vol. 1814: P. Bank, F. Baudoin, H. Föllmer, L.C.G. Rogers, M. Soner, N. Touzi, Paris-Princeton Lectures on Mathematical Finance 2002 (2003)
Vol. 1815: A. M. Vershik (Ed.), Asymptotic Combinatorics with Applications to Mathematical Physics. St. Petersburg, Russia 2001 (2003)
Vol. 1816: S. Albeverio, W. Schachermayer, M. Talagrand, Lectures on Probability Theory and Statistics. Ecole d'Eté de Probabilités de Saint-Flour XXX-2000. Editor: P. Bernard (2003)
Vol. 1817: E. Koelink, W. Van Assche (Eds.), Orthogonal Polynomials and Special Functions. Leuven 2002 (2003)
Vol. 1818: M. Bildhauer, Convex Variational Problems with Linear, nearly Linear and/or Anisotropic Growth Conditions (2003)
Vol. 1819: D. Masser, Yu. V. Nesterenko, H. P. Schlickewei, W. M. Schmidt, M. Waldschmidt, Diophantine Approximation. Cetraro, Italy 2000. Editors: F. Amoroso, U. Zannier (2003)
Vol. 1820: F. Hiai, H. Kosaki, Means of Hilbert Space Operators (2003)

Vol. 1821: S. Teufel, Adiabatic Perturbation Theory in Quantum Dynamics (2003)
Vol. 1822: S.-N. Chow, R. Conti, R. Johnson, J. Mallet-Paret, R. Nussbaum, Dynamical Systems. Cetraro, Italy 2000. Editors: J. W. Macki, P. Zecca (2003)
Vol. 1823: A. M. Anile, W. Allegretto, C. Ringhofer, Mathematical Problems in Semiconductor Physics. Cetraro, Italy 1998. Editor: A. M. Anile (2003)
Vol. 1824: J. A. Navarro González, J. B. Sancho de Salas, $\mathscr{C}^\infty$ – Differentiable Spaces (2003)
Vol. 1825: J. H. Bramble, A. Cohen, W. Dahmen, Multiscale Problems and Methods in Numerical Simulations, Martina Franca, Italy 2001. Editor: C. Canuto (2003)
Vol. 1826: K. Dohmen, Improved Bonferroni Inequalities via Abstract Tubes. Inequalities and Identities of Inclusion-Exclusion Type. VIII, 113 p, 2003.
Vol. 1827: K. M. Pilgrim, Combinations of Complex Dynamical Systems. IX, 118 p, 2003.
Vol. 1828: D. J. Green, Gröbner Bases and the Computation of Group Cohomology. XII, 138 p, 2003.
Vol. 1829: E. Altman, B. Gaujal, A. Hordijk, Discrete-Event Control of Stochastic Networks: Multimodularity and Regularity. XIV, 313 p, 2003.
Vol. 1830: M. I. Gil', Operator Functions and Localization of Spectra. XIV, 256 p, 2003.
Vol. 1831: A. Connes, J. Cuntz, E. Guentner, N. Higson, J. E. Kaminker, Noncommutative Geometry, Martina Franca, Italy 2002. Editors: S. Doplicher, L. Longo (2004)
Vol. 1832: J. Azéma, M. Émery, M. Ledoux, M. Yor (Eds.), Séminaire de Probabilités XXXVII (2003)
Vol. 1833: D.-Q. Jiang, M. Qian, M.-P. Qian, Mathematical Theory of Nonequilibrium Steady States. On the Frontier of Probability and Dynamical Systems. IX, 280 p, 2004.
Vol. 1834: Yo. Yomdin, G. Comte, Tame Geometry with Application in Smooth Analysis. VIII, 186 p, 2004.
Vol. 1835: O.T. Izhboldin, B. Kahn, N.A. Karpenko, A. Vishik, Geometric Methods in the Algebraic Theory of Quadratic Forms. Summer School, Lens, 2000. Editor: J.-P. Tignol (2004)
Vol. 1836: C. Năstăsescu, F. Van Oystaeyen, Methods of Graded Rings. XIII, 304 p, 2004.
Vol. 1837: S. Tavaré, O. Zeitouni, Lectures on Probability Theory and Statistics. Ecole d'Eté de Probabilités de Saint-Flour XXXI-2001. Editor: J. Picard (2004)
Vol. 1838: A.J. Ganesh, N.W. O'Connell, D.J. Wischik, Big Queues. XII, 254 p, 2004.
Vol. 1839: R. Gohm, Noncommutative Stationary Processes. VIII, 170 p, 2004.
Vol. 1840: B. Tsirelson, W. Werner, Lectures on Probability Theory and Statistics. Ecole d'Eté de Probabilités de Saint-Flour XXXII-2002. Editor: J. Picard (2004)
Vol. 1841: W. Reichel, Uniqueness Theorems for Variational Problems by the Method of Transformation Groups (2004)
Vol. 1842: T. Johnsen, A. L. Knutsen, $K_3$ Projective Models in Scrolls (2004)
Vol. 1843: B. Jefferies, Spectral Properties of Noncommuting Operators (2004)
Vol. 1844: K.F. Siburg, The Principle of Least Action in Geometry and Dynamics (2004)
Vol. 1845: Min Ho Lee, Mixed Automorphic Forms, Torus Bundles, and Jacobi Forms (2004)
Vol. 1846: H. Ammari, H. Kang, Reconstruction of Small Inhomogeneities from Boundary Measurements (2004)
Vol. 1847: T.R. Bielecki, T. Björk, M. Jeanblanc, M. Rutkowski, J.A. Scheinkman, W. Xiong, Paris-Princeton Lectures on Mathematical Finance 2003 (2004)
Vol. 1848: M. Abate, J. E. Fornaess, X. Huang, J. P. Rosay, A. Tumanov, Real Methods in Complex and CR Geometry, Martina Franca, Italy 2002. Editors: D. Zaitsev, G. Zampieri (2004)
Vol. 1849: Martin L. Brown, Heegner Modules and Elliptic Curves (2004)
Vol. 1850: V. D. Milman, G. Schechtman (Eds.), Geometric Aspects of Functional Analysis. Israel Seminar 2002-2003 (2004)
Vol. 1851: O. Catoni, Statistical Learning Theory and Stochastic Optimization (2004)
Vol. 1852: A.S. Kechris, B.D. Miller, Topics in Orbit Equivalence (2004)
Vol. 1853: Ch. Favre, M. Jonsson, The Valuative Tree (2004)
Vol. 1854: O. Saeki, Topology of Singular Fibers of Differential Maps (2004)
Vol. 1855: G. Da Prato, P.C. Kunstmann, I. Lasiecka, A. Lunardi, R. Schnaubelt, L. Weis, Functional Analytic Methods for Evolution Equations. Editors: M. Iannelli, R. Nagel, S. Piazzera (2004)
Vol. 1856: K. Back, T.R. Bielecki, C. Hipp, S. Peng, W. Schachermayer, Stochastic Methods in Finance, Bressanone/Brixen, Italy, 2003. Editors: M. Fritelli, W. Runggaldier (2004)
Vol. 1857: M. Émery, M. Ledoux, M. Yor (Eds.), Séminaire de Probabilités XXXVIII (2005)
Vol. 1858: A.S. Cherny, H.-J. Engelbert, Singular Stochastic Differential Equations (2005)
Vol. 1859: E. Letellier, Fourier Transforms of Invariant Functions on Finite Reductive Lie Algebras (2005)
Vol. 1860: A. Borisyuk, G.B. Ermentrout, A. Friedman, D. Terman, Tutorials in Mathematical Biosciences I. Mathematical Neurosciences (2005)
Vol. 1861: G. Benettin, J. Henrard, S. Kuksin, Hamiltonian Dynamics – Theory and Applications, Cetraro, Italy, 1999. Editor: A. Giorgilli (2005)
Vol. 1862: B. Helffer, F. Nier, Hypoelliptic Estimates and Spectral Theory for Fokker-Planck Operators and Witten Laplacians (2005)
Vol. 1863: H. Führ, Abstract Harmonic Analysis of Continuous Wavelet Transforms (2005)
Vol. 1864: K. Efstathiou, Metamorphoses of Hamiltonian Systems with Symmetries (2005)
Vol. 1865: D. Applebaum, B.V. R. Bhat, J. Kustermans, J. M. Lindsay, Quantum Independent Increment Processes I. From Classical Probability to Quantum Stochastic Calculus. Editors: M. Schürmann, U. Franz (2005)
Vol. 1866: O.E. Barndorff-Nielsen, U. Franz, R. Gohm, B. Kümmerer, S. Thorbjønsen, Quantum Independent Increment Processes II. Structure of Quantum Lévy Processes, Classical Probability, and Physics. Editors: M. Schürmann, U. Franz, (2005)
Vol. 1867: J. Sneyd (Ed.), Tutorials in Mathematical Biosciences II. Mathematical Modeling of Calcium Dynamics and Signal Transduction. (2005)
Vol. 1868: J. Jorgenson, S. Lang, $Pos_n(R)$ and Eisenstein Series. (2005)
Vol. 1869: A. Dembo, T. Funaki, Lectures on Probability Theory and Statistics. Ecole d'Eté de Probabilités de Saint-Flour XXXIII-2003. Editor: J. Picard (2005)
Vol. 1870: V.I. Gurariy, W. Lusky, Geometry of Müntz Spaces and Related Questions. (2005)

Vol. 1871: P. Constantin, G. Gallavotti, A.V. Kazhikhov, Y. Meyer, S. Ukai, Mathematical Foundation of Turbulent Viscous Flows, Martina Franca, Italy, 2003. Editors: M. Cannone, T. Miyakawa (2006)
Vol. 1872: A. Friedman (Ed.), Tutorials in Mathematical Biosciences III. Cell Cycle, Proliferation, and Cancer (2006)
Vol. 1873: R. Mansuy, M. Yor, Random Times and Enlargements of Filtrations in a Brownian Setting (2006)
Vol. 1874: M. Yor, M. Émery (Eds.), In Memoriam Paul-André Meyer - Séminaire de Probabilités XXXIX (2006)
Vol. 1875: J. Pitman, Combinatorial Stochastic Processes. Ecole d'Eté de Probabilités de Saint-Flour XXXII-2002. Editor: J. Picard (2006)
Vol. 1876: H. Herrlich, Axiom of Choice (2006)
Vol. 1877: J. Steuding, Value Distributions of $L$-Functions (2007)
Vol. 1878: R. Cerf, The Wulff Crystal in Ising and Percolation Models, Ecole d'Eté de Probabilités de Saint-Flour XXXIV-2004. Editor: Jean Picard (2006)
Vol. 1879: G. Slade, The Lace Expansion and its Applications, Ecole d'Eté de Probabilités de Saint-Flour XXXIV-2004. Editor: Jean Picard (2006)
Vol. 1880: S. Attal, A. Joye, C.-A. Pillet, Open Quantum Systems I, The Hamiltonian Approach (2006)
Vol. 1881: S. Attal, A. Joye, C.-A. Pillet, Open Quantum Systems II, The Markovian Approach (2006)
Vol. 1882: S. Attal, A. Joye, C.-A. Pillet, Open Quantum Systems III, Recent Developments (2006)
Vol. 1883: W. Van Assche, F. Marcellàn (Eds.), Orthogonal Polynomials and Special Functions, Computation and Application (2006)
Vol. 1884: N. Hayashi, E.I. Kaikina, P.I. Naumkin, I.A. Shishmarev, Asymptotics for Dissipative Nonlinear Equations (2006)
Vol. 1885: A. Telcs, The Art of Random Walks (2006)
Vol. 1886: S. Takamura, Splitting Deformations of Degenerations of Complex Curves (2006)
Vol. 1887: K. Habermann, L. Habermann, Introduction to Symplectic Dirac Operators (2006)
Vol. 1888: J. van der Hoeven, Transseries and Real Differential Algebra (2006)
Vol. 1889: G. Osipenko, Dynamical Systems, Graphs, and Algorithms (2006)
Vol. 1890: M. Bunge, J. Funk, Singular Coverings of Toposes (2006)
Vol. 1891: J.B. Friedlander, D.R. Heath-Brown, H. Iwaniec, J. Kaczorowski, Analytic Number Theory, Cetraro, Italy, 2002. Editors: A. Perelli, C. Viola (2006)
Vol. 1892: A. Baddeley, I. Bárány, R. Schneider, W. Weil, Stochastic Geometry, Martina Franca, Italy, 2004. Editor: W. Weil (2007)
Vol. 1893: H. Hanßmann, Local and Semi-Local Bifurcations in Hamiltonian Dynamical Systems, Results and Examples (2007)
Vol. 1894: C.W. Groetsch, Stable Approximate Evaluation of Unbounded Operators (2007)
Vol. 1895: L. Molnár, Selected Preserver Problems on Algebraic Structures of Linear Operators and on Function Spaces (2007)
Vol. 1896: P. Massart, Concentration Inequalities and Model Selection, Ecole d'Été de Probabilités de Saint-Flour XXXIII-2003. Editor: J. Picard (2007)
Vol. 1897: R. Doney, Fluctuation Theory for Lévy Processes, Ecole d'Été de Probabilités de Saint-Flour XXXV-2005. Editor: J. Picard (2007)
Vol. 1898: H.R. Beyer, Beyond Partial Differential Equations, On linear and Quasi-Linear Abstract Hyperbolic Evolution Equations (2007)
Vol. 1899: Séminaire de Probabilités XL. Editors: C. Donati-Martin, M. Émery, A. Rouault, C. Stricker (2007)
Vol. 1900: E. Bolthausen, A. Bovier (Eds.), Spin Glasses (2007)
Vol. 1901: O. Wittenberg, Intersections de deux quadriques et pinceaux de courbes de genre 1, Intersections of Two Quadrics and Pencils of Curves of Genus 1 (2007)
Vol. 1902: A. Isaev, Lectures on the Automorphism Groups of Kobayashi-Hyperbolic Manifolds (2007)
Vol. 1903: G. Kresin, V. Maz'ya, Sharp Real-Part Theorems (2007)
Vol. 1904: P. Giesl, Construction of Global Lyapunov Functions Using Radial Basis Functions (2007)
Vol. 1905: C. Prévôt, M. Röckner, A Concise Course on Stochastic Partial Differential Equations (2007)
Vol. 1906: T. Schuster, The Method of Approximate Inverse: Theory and Applications (2007)
Vol. 1907: M. Rasmussen, Attractivity and Bifurcation for Nonautonomous Dynamical Systems (2007)
Vol. 1908: T.J. Lyons, M. Caruana, T. Lévy, Differential Equations Driven by Rough Paths, Ecole d'Été de Probabilités de Saint-Flour XXXIV-2004 (2007)
Vol. 1909: H. Akiyoshi, M. Sakuma, M. Wada, Y. Yamashita, Punctured Torus Groups and 2-Bridge Knot Groups (I) (2007)
Vol. 1910: V.D. Milman, G. Schechtman (Eds.), Geometric Aspects of Functional Analysis. Israel Seminar 2004-2005 (2007)
Vol. 1911: A. Bressan, D. Serre, M. Williams, K. Zumbrun, Hyperbolic Systems of Balance Laws. Cetraro, Italy 2003. Editor: P. Marcati (2007)
Vol. 1912: V. Berinde, Iterative Approximation of Fixed Points (2007)
Vol. 1913: J.E. Marsden, G. Misiołek, J.-P. Ortega, M. Perlmutter, T.S. Ratiu, Hamiltonian Reduction by Stages (2007)
Vol. 1914: G. Kutyniok, Affine Density in Wavelet Analysis (2007)
Vol. 1915: T. Bıyıkoğlu, J. Leydold, P.F. Stadler, Laplacian Eigenvectors of Graphs. Perron-Frobenius and Faber-Krahn Type Theorems (2007)
Vol. 1916: C. Villani, F. Rezakhanlou, Entropy Methods for the Boltzmann Equation. Editors: F. Golse, S. Olla (2008)
Vol. 1917: I. Veselić, Existence and Regularity Properties of the Integrated Density of States of Random Schrödinger (2008)
Vol. 1918: B. Roberts, R. Schmidt, Local Newforms for GSp(4) (2007)
Vol. 1919: R.A. Carmona, I. Ekeland, A. Kohatsu-Higa, J.-M. Lasry, P.-L. Lions, H. Pham, E. Taflin, Paris-Princeton Lectures on Mathematical Finance 2004. Editors: R.A. Carmona, E. Çinlar, I. Ekeland, E. Jouini, J.A. Scheinkman, N. Touzi (2007)
Vol. 1920: S.N. Evans, Probability and Real Trees. Ecole d'Été de Probabilités de Saint-Flour XXXV-2005 (2008)
Vol. 1921: J.P. Tian, Evolution Algebras and their Applications (2008)
Vol. 1922: A. Friedman (Ed.), Tutorials in Mathematical BioSciences IV. Evolution and Ecology (2008)
Vol. 1923: J.P.N. Bishwal, Parameter Estimation in Stochastic Differential Equations (2008)

Vol. 1924: M. Wilson, Littlewood-Paley Theory and Exponential-Square Integrability (2008)
Vol. 1925: M. du Sautoy, L. Woodward, Zeta Functions of Groups and Rings (2008)
Vol. 1926: L. Barreira, V. Claudia, Stability of Nonautonomous Differential Equations (2008)
Vol. 1927: L. Ambrosio, L. Caffarelli, M.G. Crandall, L.C. Evans, N. Fusco, Calculus of Variations and Non-Linear Partial Differential Equations. Cetraro, Italy 2005. Editors: B. Dacorogna, P. Marcellini (2008)
Vol. 1928: J. Jonsson, Simplicial Complexes of Graphs (2008)
Vol. 1929: Y. Mishura, Stochastic Calculus for Fractional Brownian Motion and Related Processes (2008)
Vol. 1930: J.M. Urbano, The Method of Intrinsic Scaling. A Systematic Approach to Regularity for Degenerate and Singular PDEs (2008)
Vol. 1931: M. Cowling, E. Frenkel, M. Kashiwara, A. Valette, D.A. Vogan, Jr., N.R. Wallach, Representation Theory and Complex Analysis. Venice, Italy 2004. Editors: E.C. Tarabusi, A. D'Agnolo, M. Picardello (2008)
Vol. 1932: A.A. Agrachev, A.S. Morse, E.D. Sontag, H.J. Sussmann, V.I. Utkin, Nonlinear and Optimal Control Theory. Cetraro, Italy 2004. Editors: P. Nistri, G. Stefani (2008)
Vol. 1933: M. Petkovic, Point Estimation of Root Finding Methods (2008)
Vol. 1934: C. Donati-Martin, M. Émery, A. Rouault, C. Stricker (Eds.), Séminaire de Probabilités XLI (2008)
Vol. 1935: A. Unterberger, Alternative Pseudodifferential Analysis (2008)
Vol. 1936: P. Magal, S. Ruan (Eds.), Structured Population Models in Biology and Epidemiology (2008)
Vol. 1937: G. Capriz, P. Giovine, P.M. Mariano (Eds.), Mathematical Models of Granular Matter (2008)
Vol. 1938: D. Auroux, F. Catanese, M. Manetti, P. Seidel, B. Siebert, I. Smith, G. Tian, Symplectic 4-Manifolds and Algebraic Surfaces. Cetraro, Italy 2003. Editors: F. Catanese, G. Tian (2008)
Vol. 1939: D. Boffi, F. Brezzi, L. Demkowicz, R.G. Durán, R.S. Falk, M. Fortin, Mixed Finite Elements, Compatibility Conditions, and Applications. Cetraro, Italy 2006. Editors: D. Boffi, L. Gastaldi (2008)
Vol. 1940: J. Banasiak, V. Capasso, M.A.J. Chaplain, M. Lachowicz, J. Miękisz, Multiscale Problems in the Life Sciences. From Microscopic to Macroscopic. Będlewo, Poland 2006. Editors: V. Capasso, M. Lachowicz (2008)
Vol. 1941: S.M.J. Haran, Arithmetical Investigations. Representation Theory, Orthogonal Polynomials, and Quantum Interpolations (2008)
Vol. 1942: S. Albeverio, F. Flandoli, Y.G. Sinai, SPDE in Hydrodynamic. Recent Progress and Prospects. Cetraro, Italy 2005. Editors: G. Da Prato, M. Röckner (2008)
Vol. 1943: L.L. Bonilla (Ed.), Inverse Problems and Imaging. Martina Franca, Italy 2002 (2008)
Vol. 1944: A. Di Bartolo, G. Falcone, P. Plaumann, K. Strambach, Algebraic Groups and Lie Groups with Few Factors (2008)
Vol. 1945: F. Brauer, P. van den Driessche, J. Wu (Eds.), Mathematical Epidemiology (2008)
Vol. 1946: G. Allaire, A. Arnold, P. Degond, T.Y. Hou, Quantum Transport. Modelling, Analysis and Asymptotics. Cetraro, Italy 2006. Editors: N.B. Abdallah, G. Frosali (2008)
Vol. 1947: D. Abramovich, M. Mariño, M. Thaddeus, R. Vakil, Enumerative Invariants in Algebraic Geometry and String Theory. Cetraro, Italy 2005. Editors: K. Behrend, M. Manetti (2008)
Vol. 1948: F. Cao, J-L. Lisani, J-M. Morel, P. Musé, F. Sur, A Theory of Shape Identification (2008)
Vol. 1949: H.G. Feichtinger, B. Helffer, M.P. Lamoureux, N. Lerner, J. Toft, Pseudo-Differential Operators. Quantization and Signals. Cetraro, Italy 2006. Editors: L. Rodino, M.W. Wong (2008)
Vol. 1950: M. Bramson, Stability of Queueing Networks, Ecole d'Eté de Probabilités de Saint-Flour XXXVI-2006 (2008)
Vol. 1951: A. Moltó, J. Orihuela, S. Troyanski, M. Valdivia, A Non Linear Transfer Technique for Renorming (2008)
Vol. 1952: R. Mikhailov, I.B.S. Passi, Lower Central and Dimension Series of Groups (2008)
Vol. 1953: K. Arwini, C.T.J. Dodson, Information Geometry (2008)
Vol. 1954: P. Biane, L. Bouten, F. Cipriani, N. Konno, N. Privault, Q. Xu, Quantum Potential Theory. Editors: U. Franz, M. Schuermann (2008)
Vol. 1955: M. Bernot, V. Caselles, J.-M. Morel, Optimal Transportation Networks (2008)
Vol. 1956: C.H. Chu, Matrix Convolution Operators on Groups (2008)
Vol. 1957: A. Guionnet, On Random Matrices: Macroscopic Asymptotics, Ecole d'Eté de Probabilités de Saint-Flour XXXVI-2006 (2008)
Vol. 1958: M.C. Olsson, Compactifying Moduli Spaces for Abelian Varieties (2008)
Vol. 1959: Y. Nakkajima, A. Shiho, Weight Filtrations on Log Crystalline Cohomologies of Families of Open Smooth Varieties (2008)
Vol. 1960: J. Lipman, M. Hashimoto, Foundations of Grothendieck Duality for Diagrams of Schemes (2008)
Vol. 1961: G. Buttazzo, A. Pratelli, S. Solimini, E. Stepanov, Optimal Urban Networks via Mass Transportation (2009)

## Recent Reprints and New Editions

Vol. 1702: J. Ma, J. Yong, Forward-Backward Stochastic Differential Equations and their Applications. 1999 – Corr. 3rd printing (2007)
Vol. 830: J.A. Green, Polynomial Representations of $GL_n$, with an Appendix on Schensted Correspondence and Littelmann Paths by K. Erdmann, J.A. Green and M. Schoker 1980 – 2nd corr. and augmented edition (2007)
Vol. 1693: S. Simons, From Hahn-Banach to Monotonicity (Minimax and Monotonicity 1998) – 2nd exp. edition (2008)
Vol. 470: R.E. Bowen, Equilibrium States and the Ergodic Theory of Anosov Diffeomorphisms. With a preface by D. Ruelle. Edited by J.-R. Chazottes. 1975 – 2nd rev. edition (2008)
Vol. 523: S.A. Albeverio, R.J. Høegh-Krohn, S. Mazzucchi, Mathematical Theory of Feynman Path Integral. 1976 – 2nd corr. and enlarged edition (2008)
Vol. 1764: A. Cannas da Silva, Lectures on Symplectic Geometry 2001 – Corr. 2nd printing (2008)

*LECTURE NOTES IN MATHEMATICS* Springer

Edited by J.-M. Morel, F. Takens, B. Teissier, P.K. Maini

**Editorial Policy** (for the publication of monographs)

1. Lecture Notes aim to report new developments in all areas of mathematics and their applications - quickly, informally and at a high level. Mathematical texts analysing new developments in modelling and numerical simulation are welcome.

   Monograph manuscripts should be reasonably self-contained and rounded off. Thus they may, and often will, present not only results of the author but also related work by other people. They may be based on specialised lecture courses. Furthermore, the manuscripts should provide sufficient motivation, examples and applications. This clearly distinguishes Lecture Notes from journal articles or technical reports which normally are very concise. Articles intended for a journal but too long to be accepted by most journals, usually do not have this "lecture notes" character. For similar reasons it is unusual for doctoral theses to be accepted for the Lecture Notes series, though habilitation theses may be appropriate.

2. Manuscripts should be submitted either to Springer's mathematics editorial in Heidelberg, or to one of the series editors. In general, manuscripts will be sent out to 2 external referees for evaluation. If a decision cannot yet be reached on the basis of the first 2 reports, further referees may be contacted: The author will be informed of this. A final decision to publish can be made only on the basis of the complete manuscript, however a refereeing process leading to a preliminary decision can be based on a pre-final or incomplete manuscript. The strict minimum amount of material that will be considered should include a detailed outline describing the planned contents of each chapter, a bibliography and several sample chapters.

   Authors should be aware that incomplete or insufficiently close to final manuscripts almost always result in longer refereeing times and nevertheless unclear referees' recommendations, making further refereeing of a final draft necessary.

   Authors should also be aware that parallel submission of their manuscript to another publisher while under consideration for LNM will in general lead to immediate rejection.

3. Manuscripts should in general be submitted in English. Final manuscripts should contain at least 100 pages of mathematical text and should always include

   - a table of contents;
   - an informative introduction, with adequate motivation and perhaps some historical remarks: it should be accessible to a reader not intimately familiar with the topic treated;
   - a subject index: as a rule this is genuinely helpful for the reader.

   For evaluation purposes, manuscripts may be submitted in print or electronic form, in the latter case preferably as pdf- or zipped ps-files. Lecture Notes volumes are, as a rule, printed digitally from the authors' files. To ensure best results, authors are asked to use the LaTeX2e style files available from Springer's web-server at:

   ftp://ftp.springer.de/pub/tex/latex/svmonot1/ (for monographs).

Additional technical instructions, if necessary, are available on request from: lnm@springer.com.

4. Careful preparation of the manuscripts will help keep production time short besides ensuring satisfactory appearance of the finished book in print and online. After acceptance of the manuscript authors will be asked to prepare the final LaTeX source files (and also the corresponding dvi-, pdf- or zipped ps-file) together with the final printout made from these files. The LaTeX source files are essential for producing the full-text online version of the book (see www.springerlink.com/content/110312 for the existing online volumes of LNM).

   The actual production of a Lecture Notes volume takes approximately 12 weeks.

5. Authors receive a total of 50 free copies of their volume, but no royalties. They are entitled to a discount of 33.3% on the price of Springer books purchased for their personal use, if ordering directly from Springer.

6. Commitment to publish is made by letter of intent rather than by signing a formal contract. Springer-Verlag secures the copyright for each volume. Authors are free to reuse material contained in their LNM volumes in later publications: a brief written (or e-mail) request for formal permission is sufficient.

**Addresses:**

Professor J.-M. Morel, CMLA,
École Normale Supérieure de Cachan,
61 Avenue du Président Wilson, 94235 Cachan Cedex, France
E-mail: Jean-Michel.Morel@cmla.ens-cachan.fr

Professor F. Takens, Mathematisch Instituut,
Rijksuniversiteit Groningen, Postbus 800,
9700 AV Groningen, The Netherlands
E-mail: F.Takens@math.rug.nl

Professor B. Teissier, Institut Mathématique de Jussieu,
UMR 7586 du CNRS, Équipe "Géométrie et Dynamique",
175 rue du Chevaleret
75013 Paris, France
E-mail: teissier@math.jussieu.fr

*For the "Mathematical Biosciences Subseries" of LNM:*

Professor P.K. Maini, Center for Mathematical Biology,
Mathematical Institute, 24-29 St Giles,
Oxford OX1 3LP, UK
E-mail: maini@maths.ox.ac.uk

Springer, Mathematics Editorial I, Tiergartenstr. 17
69121 Heidelberg, Germany,
Tel.: +49 (6221) 487-8259
Fax: +49 (6221) 4876-8259
E-mail: lnm@springer.com

Printing: Krips bv, Meppel, The Netherlands
Binding: Stürtz, Würzburg, Germany